· THE OXFORD SHERLOCK HOLMES ·

General Editor
Owen Dudley Edwards

The Memoirs of
Sherlock Holmes

ARTHUR CONAN DOYLE

The Memoirs of Sherlock Holmes

Edited with an Introduction by
Christopher Roden

Oxford New York

OXFORD UNIVERSITY PRESS

Oxford University Press, Walton Street, Oxford OX2 6DP

Oxford New York
Athens Auckland Bangkok Bombay
Calcutta Cape Town Dar es Salaam Delhi
Florence Hong Kong Istanbul Karachi
Kuala Lumpur Madras Madrid Melbourne
Mexico City Nairobi Paris Singapore
Taipei Tokyo Toronto

and associated companies in
Berlin Ibadan

Oxford is a trade mark of Oxford University Press

British Library Cataloguing in Publication Data

Data available

Library of Congress Cataloging in Publication Data

Doyle, Arthur Conan, Sir, 1859-1930.
The memoirs of Sherlock Holmes / Arthur Conan Doyle ; edited with
an introduction by Christopher Roden.
p. cm. (The Oxford Sherlock Holmes)
Includes bibliographical references.
1. Holmes, Sherlock (Fictitious character)—Fiction.
2. Private investigators—England—Fiction.
3. Detective and mystery stories, English.
I. Roden, Christopher. II. Title.
III. Series : Doyle, Arthur Conan, Sir, 1859-1930.
Oxford Sherlock Holmes.
PR4622.M4 1993 823'.8—dc20 93-25375

ISBN 0-19-212309-2
ISBN 0-19-212329-7 (set)

Reprinted 1997

Printed and bound in Great Britain by
Mackays of Chatham PLC, Chatham, Kent

CONTENTS

THE MEMOIRS OF
SHERLOCK HOLMES

ACKNOWLEDGEMENTS

THE Editor of this volume would like to thank the following persons and institutions for their encouragement
and aid in its preparation: The National Library of Scotland; Catherine Cooke, Westminster Libraries, administra-
tor of the Sherlock Holmes Collection of the Marylebone
Library; the late David G. Kirby; Christopher Redmond;
the Revd Frederick J. Turner, SJ, Archivist, Stonyhurst
College, Lancashire; Cameron Hollyer, formerly custodian
of the Arthur Conan Doyle Collection, the Metropolitan
Toronto Reference Library, Toronto, Ontario, Canada, and
the Library itself; the Manuscripts Division, Tracy W.
McGregor Library, Special Collections Department,
University of Virginia Library, Charlottesville, Virginia,
USA; the William Hill Organization; Alfred Dunhill Ltd;
Douglas S. Warren; the Stormy Petrels of British Columbia,
Canada.

And, above all, Barbara Roden.

The General Editor's thanks for obligations incurred by
the entire series are asserted in the acknowledgements to *A
Study in Scarlet*.

GENERAL EDITOR'S PREFACE
TO THE SERIES

ARTHUR CONAN DOYLE told his *Strand* editor, Herbert Greenhough Smith (1855–1935), that 'A story always comes to me as an organic thing and I never can recast it without the Life going out of it.'[1]

On the whole, this certainly seems to describe Conan Doyle's method with the Sherlock Holmes stories, long and short. Such manuscript evidence as survives (approximately half the stories) generally bears this out: there is remarkably little revision. Sketches or scenarios are another matter. Conan Doyle was no more bound by these at the end of his literary life than at the beginning, whence scraps of paper survive to tell us of 221B Upper Baker Street where lived Ormond Sacker and J. Sherrinford Holmes. But very little such evidence is currently available for analysis.

Conan Doyle's relationship with his most famous creation was far from the silly label 'The Man Who Hated Sherlock Holmes': equally, there was no indulgence in it. Though the somewhat too liberal Puritan Micah Clarke was perhaps dearer to him than Holmes, Micah proved unable to sustain a sequel to the eponymous novel of 1889. By contrast, 'Sherlock' (as his creator irreverently alluded to him when not creating him) proved his capacity for renewal 59 times (which Conan Doyle called 'a striking example of the patience and loyalty of the British public'). He dropped Holmes in 1893, apparently into the Reichenbach Falls, as a matter of literary integrity: he did not intend to be written off as 'the Holmes man'. But public clamour turned Holmes into an economic asset that could not be ignored. Even so, Conan Doyle could not have continued to write about

---

[1] Undated letter, quoted by Cameron Hollyer, 'Author to Editor', *ACD— The Journal of the Arthur Conan Doyle Society*, 3 (1992), 19–20. Conan Doyle's remark was probably *à propos* 'The Red Circle' (*His Last Bow*).

Holmes without taking some pleasure in the activity, or indeed without becoming quietly proud of him.

Such Sherlock Holmes manuscripts as survive are frequently in private keeping, and very few have remained in Britain. In this series we have made the most of two recent facsimiles, of 'The Dying Detective' and 'The Lion's Mane'. In general, manuscript evidence shows Conan Doyle consistently underpunctuating, and to show the implications of this 'The Dying Detective' (*His Last Bow*) has been printed from the manuscript. 'The Lion's Mane', however, offers the one case known to us of drastic alterations in the surviving manuscript, from which it is clear from deletions that the story was entirely altered, and Holmes's role transformed, in the process of its creation.

Given Conan Doyle's general lack of close supervision of the Holmes texts, it is not always easy to determine his final wishes. In one case, it is clear that 'His Last Bow', as a deliberate contribution to war propaganda, underwent a ruthless revision at proof stage—although (as we note for the first time) this was carried out on the magazine text and lost when published in book form. But nothing comparable exists elsewhere.

In general, American texts of the stories are closer to the magazine texts than British book texts. Textual discrepancies, in many instances, may simply result from the conflicts of sub-editors. Undoubtedly, Conan Doyle did some re-reading, especially when returning to Holmes after an absence; but on the whole he showed little interest in the constitution of his texts. In his correspondence with editors he seldom alluded to proofs, discouraged ideas for revision, and raised few—if any—objections to editorial changes. For instance, we know that the *Strand*'s preference for 'Halloa' was not Conan Doyle's original usage, and in this case we have restored the original orthography. On the other hand, we also know that the *Strand* texts consistently eliminated anything (mostly expletives) of an apparently blasphemous character, but in the absence of manuscript confirmation we have normally been unable to restore what were probably stronger original versions. (In any case, it is perfectly possible that Conan Doyle, the consummate professional, may have come to exercise self-censorship in the certain knowledge that editorial changes would be imposed.)

Throughout the series we have corrected any obvious errors, though these are comparatively few: the instances are at all times noted. (For a medical man, Conan Doyle's handwriting was commendably legible, though his 'o' could look like an 'a'.) Regarding the order of individual stories, internal evidence makes it clear that 'A Case of Identity' (*Adventures*) was written before 'The Red-Headed League' and was intended to be so printed; but the 'League' was the stronger story and the *Strand*, in its own infancy, may have wanted the series of Holmes stories established as quickly as possible (at this point the future of both the Holmes series and the magazine was uncertain). Surviving letters show that the composition of 'The Solitary Cyclist' (*Return*) preceded that of 'The Dancing Men' (with the exception of the former's first paragraph, which was rewritten later); consequently, the order of these stories has been reversed. Similarly, the stories in *His Last Bow* and *The Case-Book of Sherlock Holmes* have been rearranged in their original order of publication, which—as far as is known—reflects the order of composition. The intention has been to allow readers to follow the fictional evolution of Sherlock Holmes over the forty years of his existence.

The one exception to this principle will be found in *His Last Bow*, where the final and eponymous story was actually written and published after *The Valley of Fear*, which takes its place in the Holmes canon directly after the magazine publication of the other stories in *His Last Bow*; but the removal of the title story to the beginning of the *Case-Book* would have been too radically pedantic and would have made *His Last Bow* ludicrously short. Readers will note that we have already reduced the extent of *His Last Bow* by returning 'The Cardboard Box' to its original location in the *Memoirs of Sherlock Holmes* (after 'Silver Blaze' and before 'The Yellow Face'). The removal of 'The Cardboard Box' from the original sequence led to the inclusion of its introductory passage in 'The Resident Patient': this, too, has been returned to its original position and the proper opening of 'The Resident Patient' restored. Generally, texts have been derived from first book publication collated with magazine texts and, where possible, manuscripts; in the case of 'The Cardboard Box' and 'The Resident Patient', however, we have employed the *Strand* texts, partly because of the restoration of the latter's opening, partly to give readers a flavour of the magazine in which the Holmes stories made their first, vital conquests.

In all textual decisions the overriding desire has been to meet the author's wishes, so far as these can be legitimately ascertained from documentary evidence or application of the rule of reason.

One final plea. If you come to these stories for the first time, proceed now to the texts themselves, putting the introductions and explanatory notes temporarily aside. Our introductions are not meant to introduce: Dr Watson will perform that duty, and no one could do it better. Then, when you have mastered the stories, and they have mastered you, come back to us.

OWEN DUDLEY EDWARDS

University of Edinburgh

INTRODUCTION

S IR ARTHUR CONAN DOYLE (1859–1930) wrote sixty stories featuring his detective creation Sherlock Holmes which, with the exception of *A Study in Scarlet* (1887) and *The Sign of the Four* (1890), were all published in the *Strand Magazine*. Conan Doyle's first contribution to the *Strand* was 'The Voice of Science', a short story published anonymously (although credited in the magazine's index) in March 1891. His final story, 'The Last Resource', appeared in December 1930, some months after his death. The fortunes of the *Strand* waxed and waned with Conan Doyle's contributions. A new series of Sherlock Holmes stories was a virtual guarantee that there would be a significant surge in demand for the already popular monthly: the stories that were later collectively published as *The Memoirs of Sherlock Holmes* would make their own particular impact on the magazine's circulation following the December 1892 issue. Conan Doyle remained loyal to the *Strand*, and the *Strand* to Conan Doyle, for a period of thirty-nine years. Their mutual success during that period was due, in no small measure, to the foresight of the *Strand*'s first editor, Herbert Greenhough Smith (1855–1935), who recognized the potential of Conan Doyle's short stories. In 'The Passing of Conan Doyle' (*Strand*, September 1930), Greenhough Smith wrote:

> It was in 1891 that, as Editor of *The Strand Magazine*, I received the first of these stories which were destined to become famous over all the world as 'The Adventures of Sherlock Holmes'. I have cause to remember the occasion well. *The Strand Magazine* was in its infancy in those days: good story-writers were scarce, and here to an editor, jaded with wading through reams of impossible stuff, comes a gift from Heaven, a godsend in the shape of a story that brought a gleam of happiness into the despairing life of this weary editor. Here was a new and gifted story-writer; there was no mistaking the ingenuity of plot, the limpid clearness of style, the perfect art of telling a story.

That story, 'A Scandal in Bohemia', was the first of the Sherlock Holmes short stories to appear in the *Strand* (July 1891), and also the first of the series later collected as *The Adventures of Sherlock Holmes* (1892). 'A Scandal in Bohemia' had been submitted by Conan Doyle's agent, A(lexander) P(ollock) Watt (1834–1914), during the author's absence in Vienna. Conan Doyle had left his former Southsea medical practice in December 1890 and travelled with his wife to Vienna, where he proposed to study eye medicine. His opinion of the time he spent in Vienna was that he could certainly have learned far more in London, but he did manage to see a little of Viennese society. To assist in covering expenses, Conan Doyle wrote a short novel during his time in the city. *The Doings of Raffles Haw* (1892) is generally overlooked today as being of little consequence. Nevertheless, Conan Doyle used the interesting plot device of alchemy, suggesting that he may have been interested in the activities of the Rosicrucian movement at the time. The alchemical practices of the Rosicrucians had also featured as a plot device in his much earlier story 'The Silver Hatchet' (1883). Such interest should not be surprising in view of the wide examination of religions and philosophies which he revealed in his semi-autobiographical novel *The Stark Munro Letters* (1895)—an examination of beliefs brought about by his rejection of the Roman Catholicism into which he had been born and schooled. A further point of interest is that Conan Doyle's brother-in-law, Ernest William Hornung (1866–1921), was to adopt the name Raffles as a surname for the chief character in his highly successful series of adventures *The Amateur Cracksman* (1899), and also gave him Arthur Conan Doyle's first name. Hornung's dedication to that series read: 'To ACD/This form of flattery'.

Following his return to London in early 1891, Conan Doyle practised unsuccessfully as an oculist. In August of that year, following a particularly virulent attack of influenza, he made the decision to 'live by the pen'. His ambition for some time had been to write historical novels in the style of Scott and Macaulay, both of whom he admired immensely.

Micah Clarke (1889) had been his first effort, and the success of that novel had stimulated interest in Conan Doyle's other work, including the two Sherlock Holmes stories *A Study in Scarlet* and *The Sign of the Four*. His next major historical novel, *The White Company*, was published in October 1891, followed in October 1892 by his first Napoleonic novel, *The Great Shadow*, which had been written between April and June 1892.

Conan Doyle's popularity as a novelist was increasing and, whilst this alone was sufficient cause for celebration, November 1892 brought an event of further happiness for himself and his first wife, Louise: the birth of their second child, a son, Alleyne Kingsley. The choice of Alleyne for a name acknowledged ACD's fondness for the character Alleyne Edricson of *The White Company*.

The *Strand*, meanwhile, had been pressing for further Sherlock Holmes stories: the first series of six adventures had been very successful with the public, and the magazine's fortunes were going from strength to strength. Despite this, Conan Doyle was not convinced that this was work upon which he should be engaged. In November 1891, following completion of 'The Adventure of the Beryl Coronet', he wrote to his mother:

I have done five of the Sherlock Holmes stories of the new series. . . . I think that they are up to the standard of the first series, & the twelve ought to make a rather good book of the sort. I think of slaying Holmes in the sixth & winding him up for good & all. He takes my mind from better things. I think your golden haired idea has the making of a tale in it, but I think it would be better not as a detective tale, but as a separate one.

Mrs Doyle was horrified and, fortunately, her protests won a reprieve for the popular detective. Her idea was used, but Conan Doyle replaced the golden hair with the chestnut locks of Miss Violet Hunter and introduced Sherlock Holmes into the plot to produce 'The Copper Beeches'. The death of Sherlock Holmes had been averted and, had the public known, there would have been a huge sigh of relief.

The relief and the reprieve were, however, to be only temporary.

Conan Doyle had first been approached for a further series of Holmes stories in February 1892, when he was working hard on his latest novel, *The Refugees*, which would be published in May 1893. He had also, on the suggestion of James Matthew Barrie (1860–1937), been working on a dramatization of his earlier short story 'A Straggler of '15', and the play he produced came to be known as *Waterloo*. It so impressed Henry Irving that he bought the full acting rights and proceeded to endow the role of Brewster with his own distinctive style.

But Conan Doyle was not keen to produce further Holmes stories: the intricacies of plot necessary for the short detective story were as time-consuming as those for a novel. In *Memories and Adventures* (1924) he wrote 'The difficulty of the Holmes work was that every story needed as clear-cut and original a plot as a longish book would do. One cannot without effort spin plots at such a rate. They are apt to become thin or to break.' He considered how he could get around the *Strand*'s latest approach and decided to ask one thousand pounds for the new series in the hope that this would prove a deterrent. (He had been paid thirty guineas for each of the stories in the first series, and fifty guineas for each of the second series.) But the *Strand* accepted his terms without hesitation and Conan Doyle had now to begin considering ideas for the stories that would eventually become *The Memoirs of Sherlock Holmes*.

However, not everything Conan Doyle touched was proving quite as successful as Sherlock Holmes. In September 1892 he received a telegram from Barrie which resulted in his journeying to Aldeburgh, where Barrie sought his assistance with the libretto of a light opera. *Jane Annie; or, The Good Conduct Prize* opened at the Savoy Theatre on 13 May 1893 and ran until 1 July. ACD wrote, 'I collaborated with him on this, but in spite of our joint efforts, the piece fell flat'. Conan Doyle and Barrie remained good-humoured, and the failure inspired Barrie to write one of the earliest parodies of Sherlock Holmes, 'The Adventure of the Two Collaborators', which ACD enjoyed immensely and reproduced in *Memories and Adventures* (see Appendix 1).

The first of the new series of Holmes stories, 'Silver Blaze', appeared in the *Strand* in December 1892. When Harry How visited Conan Doyle in June 1892 to interview him for the *Strand*, he was able to report:

... but he [ACD] declares that already he has enough material to carry him through another series, and merrily assures me that he thought the opening story of the next series of 'Sherlock Holmes', ... was of such an unsolvable character, that he had positively bet his wife a shilling that she would not guess the true solution of it until she got to the end of the chapter.

That Conan Doyle was tiring of Holmes is, however, very apparent. Although his mother had succeeded in delaying the detective's death, disposing of Holmes was still very much in his mind. J. M. Barrie, writing in his autobiography *The Greenwood Hat* (1937), mentioned that he sat with Conan Doyle on the sea-shore at Aldeburgh when the decision was made to kill Sherlock Holmes. Whether this occurred when the two met to discuss *Jane Annie* in September 1892 is not clear. There are varying accounts which relate ACD's decision to kill Holmes: Richard Lancelyn Green, in *The Uncollected Sherlock Holmes* (1983), identifies a visit ACD made to Switzerland in August 1893 when he would have visited the Reichenbach Falls, as the occasion when he made the decision. Whenever it was, the decision was seemingly irrevocable, and on 6 April 1893, ACD wrote to his mother:

My Dearest Mam,
All is very well down here. I am in the middle of the last Holmes story, after which the gentleman vanishes, never never to reappear. I am weary of his name. The medical stories also are nearly finished, so I shall have a clear sheet soon. Then for playwriting and lecturing for a couple of years ...

There was no going back. Conan Doyle recalled his decision in *Memories and Adventures*:

... after I had done two series ... I saw that I was in danger of having my hand forced, and of being entirely identified with what I regarded as a lower stratum of literary achievement. Therefore as a sign of my resolution I determined to end the life of my hero. The idea was in my mind when I went with my wife for a short holiday in Switzerland, in the course of which we saw the wonderful falls of Reichenbach, a terrible place, and one that I thought would make a worthy tomb for poor Sherlock, even if I buried my banking account along with him ...

Christmas 1893 was to be a sad one for readers of the *Strand* and admirers of Sherlock Holmes. The December edition of the magazine included 'The Adventure of The Final Problem', which was prefaced with Sidney Paget's illustration of Sherlock Holmes and Professor Moriarty locked in combat above the Reichenbach Falls and captioned 'The Death of Sherlock Holmes'. It is said that the effect on the readership was so profound that young City men put mourning crepe on their silk hats. One anguished correspondent wrote directly to Conan Doyle: 'You brute!'. George Newnes, who had founded the *Strand* in 1891, referred to Sherlock Holmes's death as 'a dreadful event'. Well he might: the *Strand*'s circulation had been built on the back of the success of Sherlock Holmes. Who knew what the future might bring?

Conan Doyle was suffering his own particular grief at the time 'The Final Problem' appeared: his father, Charles Altamont Doyle, who had for many years been institutionalized, had died on 10 October 1893. The inevitability of death entered ACD's life in 1893: and it was that inevitability, as well as the human inability to escape from past wrongdoings, that were reflected in *The Memoirs of Sherlock Holmes*.

The plots of both *A Study in Scarlet* and *The Sign of the Four* were concerned with retribution: one for the suffering caused by Mormon polygamy, the other for a betrayal and double-cross which had taken place in colonial India. Conan Doyle used the idea again when he came to write the stories that were published as the *Adventures* (for example in

'The Boscombe Valley Mystery'). In the *Memoirs*, however, the theme of wrongdoings is not only continued, it is intensified in some of the stories by the implication and actuality of infidelity, a theme which, for Conan Doyle, was something of a departure from the more simplistic approach of the earlier Holmes short stories.

Nothing specific in Conan Doyle's background suggests a reason for the sudden inclusion of infidelity as a plot device in his stories, and his interest in the theme would seem to be unrelated to his interest in Divorce Law Reform, with which he was not to become actively involved until 1906. There was, however, the problem of his father's alcoholism, which may have been the cause of earlier unpleasant scenes in the Doyle household. Conan Doyle was, without doubt, much closer to his mother than to his father, and it is conceivable that the onset of Charles Doyle's epilepsy following his institutionalization in the early 1880s caused ACD to consider the past problems within the family, thereby providing the trigger for this particular theme. Unpleasant scenes do not, of course, imply infidelity, but ACD could have been viewing the problem of the 'wronged woman' in general, and it was in that light that he almost certainly viewed his mother's own predicament. When considering the problem in that manner, infidelity becomes a simple step in the thought process.

Conan Doyle has been alternately described as a straightforward and a complex personality. Certainly, if the themes of his many short stories are considered, his was a complex character. His semi-autobiographical novel *The Stark Munro Letters* (1895) indicates the turmoil of his own mind on the subject of religion and destiny. At about the time that novel was written, he was describing the various battles within his mind in his poem 'The Inner Room':

> It is mine—the little chamber,
>   Mine alone.
> I had it from my forebears
>   Years agone.
> Yet within its walls I see

A most motley company,
And they one and all claim me
    As their own.

There's one who is a soldier
    Bluff and keen;
Single-minded, heavy fisted,
    Rude of mien.
He would gain a purse or stake it,
He would win a heart or break it,
He would give a life or take it,
    Conscience-clean.

And near him is a priest
    Still schism-whole;
He loves the censer-reek
    And organ-roll.
He has leanings to the mystic,
Sacramental, eucharistic;
And dim yearnings altruistic
    Thrill his soul.

There's another who with doubts
    Is overcast;
I think him younger brother
    To the last.
Walking wary stride by stride,
Peering forwards anxious-eyed,
Since he learned to doubt his guide
    In the past.

And 'mid them all, alert,
    But somewhat cowed,
There sits a stark-faced fellow,
    Beetle-browed,
Whose black soul shrinks away
From a lawyer-ridden day,
And has thoughts he dare not say
    Half avowed.

There are others who are sitting,
    Grim as doom,

In the dim ill-boding shadow
    of my room.
Darkling figures, stern or quaint,
Now a savage, now a saint,
Showing fitfully and faint
    Through the gloom.

And those shadows are so dense,
    There may be
Many—very many—more
    Than I see.
They are sitting day and night
Soldier, rogue, and anchorite;
And they wrangle and they fight
    Over me.

If the stark-faced fellow win,
    All is o'er!
If the priest should gain his will,
    I doubt no more!
But if each shall have his day,
I shall swing and I shall sway
In the same old weary way
    As before.

Conan Doyle's conception of infidelity is shown in three of the stories of the *Memoirs*, and each example is very different from the others. In 'Silver Blaze', John Straker is indulging the very expensive tastes of a lady who is not his wife. He is also deceiving his wife by using the alias of William Darbyshire, whom she believes to be a friend using the marital home's address for the convenience of receiving some of his mail. Straker's indulgences lead him to the actions that are to be the cause of his death. Holmes refrains from passing judgement as he summarizes the situation:

... Straker was leading a double life, and keeping a second establishment. The nature of the bill showed that there was a lady in the case, and one who had expensive tastes ... I have no doubt that this woman had plunged him over head and ears in debt, and so led him into this miserable plot.

Michael Harrison, the eminent Sherlockian scholar, ex-
pounded an interesting theory for the source of 'Silver
Blaze' (*Immortal Sleuth* (1983)). His thesis, briefly, was that
Conan Doyle became interested in the notorious Mordaunt
scandal of 1869. The case concerned a divorce action
brought about by the birth of a son who suffered *ophthalmia
neonatorum*: gonorrheal infection of the eyes. Conan Doyle's
interest, Harrison suggests, could have been either through
his studies of venereal disease (the subject of his doctoral
thesis), or through his study of eye medicine a few years
prior to writing 'Silver Blaze'. The connection Harrison
makes is that the leading figure in the case, Sir Frederick
Johnstone, had himself been a winning racehorse owner and
had, in 1881, brought off the 'classic hat-trick', i.e., winning
the Derby, the Oaks, and the Ascot Gold Cup. The name
of the racehorse achieving this feat was none other than St
Blaise. Another prominent figure involved in the Mordaunt
scandal was Albert Edward, Prince of Wales, who appears
in the story disguised as the owner of one of the horses
competing for the Wessex Cup. Conan Doyle conceals the
Prince's identity by using the lesser-known title, Duke of
Balmoral.

When *The Memoirs of Sherlock Holmes* was published in book
form as volume three of *The Strand Library* series on 13
December 1893, only eleven of the twelve stories which had
previously been serialized in the *Strand* were included. 'The
Adventure of the Cardboard Box' was omitted and a great
deal of speculation has taken place over the reason why
Conan Doyle should have wished this particular story to b
suppressed. ACD himself appears to have offered differing
excuses: that it was out of place in a collection designed for
boys; that it was more sensational than he cared for; and
that it was a weak story. Whether one can consider a
collection of stories which had appeared in the *Strand* as
being intended for boys' consumption is open to debate. It
seems an unrealistic explanation when one considers some
of the other adventures: 'A Scandal in Bohemia' (*Adventures*),

'The Yellow Face', 'Silver Blaze' (*Memoirs*), 'The Solitary Cyclist', and 'Charles Augustus Milverton' (*Return*) all have features which might then have been considered to render them unsuitable for boys' reading material. 'The Cardboard Box' is certainly not a weak story, and Conan Doyle's own regard for its opening paragraphs, which he transposed into 'The Resident Patient', seems to refute his statement. One is led to the conclusion, therefore, that something of which we are unaware, or can only speculate upon, happened in Conan Doyle's life between the time the story appeared in the *Strand* and the time the collection was published in book form. Whatever it was that led to the suppression of 'The Cardboard Box' was apparently expunged from Conan Doyle's memory, or no longer held prohibitive weight with him, by the time the collection of stories entitled *His Last Bow* was published in 1917, as the story was included in that volume.

'The Cardboard Box' has murder and implied adultery as its theme, and this double horror may be the true reason for Conan Doyle's original suppression of the story. The narrative, however, is anything but explicit on the subject of adultery. Certainly there is an illicit love affair between Mary, Jim Browner's wife, and Alec Fairbairn; but Browner's narrative is circumspect regarding its details. It should be remembered that 'Silver Blaze' also discloses an illicit love affair. It is tempting to draw the conclusion that Conan Doyle may have been influenced in his decision to withdraw the story by sensitivity within his family over the question of violence brought about by alcoholism. The picture he paints of Browner tells its own story:

> I had been drinking hard of late, and the two things together fairly turned my brain. There's something throbbing in my head now, like a docker's hammer, but that morning I seemd to have all Niagara whizzing and buzzing in my ears ... I was like a wild beast that had tasted blood ...

Browner is a gentle soul who has been driven back to drink by jealousy and the threat to his marriage caused by the manœuvrings of his devious and frustrated sister-in-law. The influence Sarah Cushing has exerted over Browner is almost vampirical and she may be compared to the evil Miss Penelosa of Conan Doyle's *The Parasite* (1894), a story which would have been conceived at about the same time as 'The Cardboard Box' was written. We see a reflective and philosophical Conan Doyle expressing himself in the final paragraphs of the story, when Holmes says:

What is the meaning of it, Watson? . . . What object is served by this circle of misery and violence and fear? It must tend to some end, or else our universe is ruled by chance, which is unthinkable. But what end? There is the great standing perennial problem to which human reason is as far from an answer as ever.

The third story of the collection, 'The Yellow Face', introduces infidelity only as a suspicion in the mind of one of the story's chief characters, Grant Munro. Munro and his wife have been happily married for three years when he suddenly begins, because of a simple request for money, to mistrust her: '. . . I find that there is something in her life and in her thoughts of which I know as little as if she were the woman who brushes by me in the street. We are estranged, and I want to know why.' Holmes, after hearing Munro's account, jumps to an erroneous conclusion: that Munro's wife is harbouring her first husband and is, therefore, guilty of bigamy. There is, in fact, a far more plausible and innocent explanation which is discovered by Munro himself. 'The Yellow Face' is the third of Conan Doyle's stories to have been inspired by the black anti-slavery leader, Henry Highland Garnet (1815–1882), whom Conan Doyle attended in his capacity as ship's doctor when Garnet, then United States Minister to Liberia, joined the *Mayumba* for three days early in 1882, a short time before his death. The two other stories, 'J. Habakuk Jephson's Statement' (1884), and 'The Five Orange Pips' (*Adventures*), have as their primal cause the murders of blacks by whites in the American South, the first under slavery, the second by the Ku Klux Klan during Reconstruction. It seems likely that, as originally conceived,

'The Yellow Face' assumed that John Hebron was murdered in Atlanta, possibly by being burned to death in a conflagration which also devoured his legal office and papers. During American Reconstruction the hope of black–white equality animated many Southern blacks and supportive whites, and Republican support for black enfranchisement and participation in the state's public life could well have encouraged a black of intellectual distinction and courage to open up a legal practice in a black-dominated, poverty-stricken part of Atlanta, which was rebuilding itself after the Civil War fire. The passage of the Fifteenth Amendment to the United States Constitution in 1870, stating that no person should have his right to vote abridged by reason of race, colour, or previous condition of servitude, might naturally suggest that a healthy future existed for racial integration. The Georgia Constitution of 1865 had explicitly asserted that 'The marriage relation between white persons and persons of African descent is forever prohibited, and such marriages shall be null and void.' (Art. 5, sect. 1, para. 9). But such a social revolutionary as Hebron might well contemplate bringing suit against the State of Georgia under the Constitution, declaring this assertion to be an abridgement of his liberties under the 15th, 14th, and other Amendments in the hope of having it struck down by the United States Supreme Court, and thus legalizing future intermarriage.

All such hopes would have disappeared after the pulling out of the last Federal troops from the South (from South Carolina, Louisiana, and Florida) in 1877. Conan Doyle would have known from the press in 1877 about that; he may not have realized that the Republican pro-black Reconstruction in Georgia fell as early as 1872, which would have made any continued existence for a marriage of black man and white woman in Atlanta impossible. It seems clear that when Hebron died he was maintaining his wife and child in New York or another Northern state, and that he had returned to Atlanta alone. 1878, as the year after Reconstruction ended, seems a probable presumed date of death of Hebron by Conan Doyle, especially with a huge yellow-fever epidemic raging in the South that year. As originally conceived, then, the story may have involved Mrs Grant Munro telling her husband that Hebron had died in 1878 of yellow fever, not that he had been lynched in Atlanta; and the dénouement would have revealed the truth. But this would have involved a long digression on the recent history of Georgia, diminishing the dramatic effect of discovery and decision by Grant Munro. As it is, the text is not incompatible with the lynching hypothesis, which, tragically, would have been by far the most probable end of Hebron's story.

The effect Garnet had on Conan Doyle seems to have been immense. The sympathy with which the dénouement of 'The Yellow Face' is handled contrasts sharply with some of Conan Doyle's earlier writings concerning blacks. In an article entitled 'On the Slave Coast with a Camera', written for *The British Journal of Photography* (vol. 29, 31 March, 7 April 1882, pp. 185–7, 202–3), he wrote:

A great deal has been said about the regeneration of our black brothers and the latent virtues of the swarthy races. My own experience is that you abhor them on first meeting them, and gradually learn to dislike them a great deal more as you become better acquainted with them . . .

Of Henry Highland Garnet, Conan Doyle wrote 'This negro gentleman did me good, for a man's brain is an organ for the formation of his own thoughts and also for the digestion of other people's, and it needs fresh fodder' (*Memories and Adventures*). That Conan Doyle was able to display such sensitivity in handling the subject of inter-racial marriage which, for its time, was a taboo subject, shows that, despite earlier prejudices, he had without doubt digested some of the fodder placed before him by Garnet.

'The Yellow Face' has many elements that had come to be expected in a Sherlock Holmes story; but it also has some that were not. In typical style, deriving from Edgar Allan Poe's similarly accomplished detective C. Auguste Dupin in 'The Murders in the Rue Morgue', Holmes brilliantly deduces the habits of the owner of a pipe left behind in his Baker Street rooms. This is quintessential Sherlock Holmes; but the scene contrasts strongly with the uncharacteristic errors that Conan Doyle forces his hero to make later in the story. The revelation of fallibility strikes a new and surprising note.

The first three stories of the *Memoirs* broach the subject of marital infidelity; 'The Musgrave Ritual' shows a relationship between two lovers in which the infidelity of one, coupled with a greed to acquire something belonging to someone else, leads to eventual death. Brunton, the butler, is engaged to Rachel Howells, the second housemaid. His Don Juan temperament has, however, led him into an entanglement with Janet Tregellis, the daughter of the head gamekeeper. The effect on Rachel Howells, a good girl but of excitable Welsh temperament, is devastating and she seizes the first opportunity to take revenge for her rejection. It is interesting that Conan Doyle should have chosen two girls of Celtic stock to be rivals in this story: it was certainly a powerful means of implying the fiery rivalry that would have existed between the two.

The entanglement between Brunton and Rachel Howells is kept apart from the main adventure, the puzzle of the Musgrave ritual, and was perhaps used by Conan Doyle merely to add length and complication to the story. Certainly the sub-plot is not satisfactorily resolved: Brunton dies, but Rachel Howells disappears, and Holmes shows no interest in tracing the whereabouts of one of the few women killers he encounters in his long career. Feminist critics have noted that Holmes sees women as intrinsically destructive of the logical system by which he operates; Rachel Howells has been the agent of destruction of a logician working by the methods he later uses himself. That Brunton usurps the legacy and meaning of the Musgraves worries him not at all.

For Sherlockians, 'The Musgrave Ritual' offers glimpses of Holmes's domestic habits. It is the second story to reveal details of his time at university (though in doing so has added to the body of speculation on the location of Holmes's *alma mater*). It also offers a list (again, tantalizing both for his contemporary readers and for subsequent Sherlockians) of Holmes's early cases: the Tarleton murders; the case of Vamberry, the wine merchant; the adventure of the old Russian woman; the singular affair of the aluminium crutch; Ricoletti of the club foot and his abominable wife. As Conan Doyle wrote in *Memories and Adventures*, 'Heaven knows how many titles I have thrown about in a casual way, and how many readers have begged me to satisfy their curiosity.'

The historical aspects of 'The Musgrave Ritual' are not truly an example of past misdeeds, a theme Conan Doyle uses with particular effect in three stories of the *Memoirs*: 'The "Gloria Scott" ', 'The Crooked Man', and 'The Resident Patient'. The theme is first encountered in the *Memoirs* in 'The "Gloria Scott" ' when a young Sherlock Holmes is called in by a college friend, Victor Trevor, following the death of Trevor's father. Holmes had already made Trevor senior's acquaintance and had deduced circumstances in his past that had caused some consternation. The story revolves around murder, betrayal, and changed identities from the days of transportation to Australia, and has similarities with 'The Boscombe Valley Mystery' (*Adventures*).

The second story of revenge/past misdeeds, 'The Crooked Man', is much stronger. Interwoven into the plot is a powerful story of thwarted love and mutilation. There is possibly no more horrible a story in the Holmes cycle: the betrayal, torture, and eventual enslavement Henry Wood has to suffer portray a nightmarish existence and demand that he be given both Holmes's and the reader's sympathy. Holmes obliges by allowing Wood to go free following the death of Colonel Barclay, even though there can be little doubt that Wood was the cause of it. 'The Crooked Man' draws on *The Sign of the Four* for its theme of revenge, and also has the presence of a small unidentified companion: a mongoose, as opposed to Tonga the Andaman Islander in *The Sign of the Four*. But the mongoose is not intended to create the atmosphere of the unknown which is expected and achieved with Tonga. Henry Wood, like Jonathan Small, has been cheated—Small of material treasure, Wood of the woman he loved. Both leave India, after a period of imprisonment, with permanent physical disabilities; but Wood's injuries and disability are far more horrific than those of Small. 'The Crooked Man' also borrows something from the plot of *A Study in Scarlet*—the plight of a man whose woman has been stolen from him by someone prepared to abuse his greater authority.

The colonial background featured strongly in many of Conan Doyle's earlier stories: 'The Mystery of Sasassa Valley' (1879), 'Bones: The April Fool of Harvey's Sluice' (1882), 'The Gully of Bluemansdyke' (1881), and 'The Parson of Jackman's Gulch' (1885) are all examples. One story seems to have been particularly influential on 'The Resident Patient'. 'My Friend the Murderer' (1882) concerns a criminal who has obtained immunity from prosecution by giving evidence against his fellows. As Owen Dudley Edwards has discussed,[1] the plot draws on the careers and prosecution of the notorious corpse-selling murderers, Burke and Hare (who are mentioned in the story). In 'The Resident Patient' Dr Percy Trevelyan is set up in practice by a Mr Blessington, who turns out to be the member of the Worthingdon bank gang who turned informer, thereby escaping prosecution. The betrayal is avenged when the other members of the gang are released from prison and visit the resident patient, Blessington, for the purpose of meting out their own form of justice.

There are further minor themes running through the stories of the *Memoirs*: espionage features in 'The Naval Treaty'; avarice in 'The Reigate Squire' and 'The Greek Interpreter'. 'The Stockbroker's Clerk' is a tale of duping and borrows heavily from the idea Conan Doyle used in 'The Red-Headed

[1] O. D. Edwards, *The Quest for Sherlock Holmes* (Edinburgh: Mainstream, 1983), 195–6.

League' (*Adventures*) and which he was to re-work for a second time in 'The Three Garridebs' (*Casebook*).

The problem with which Conan Doyle was grappling throughout the writing of this series of stories, the manner of Holmes's demise, is at last confronted in 'The Final Problem', in which he provides Holmes with his most potent adversary. In Professor Moriarty, Conan Doyle created a powerful embodiment of malevolence and one of the most memorable villains of fiction. Moriarty is the epitome of all that is evil, opposing in every respect the portrayal of Holmes presented in the twenty-five previous stories. The respect Holmes has for the intellectual abilities of his opponent is apparent in his description of Moriarty:

He is the Napoleon of crime, Watson. He is the organizer of half that is evil and of nearly all that is undetected in this great city. He is a genius, a philosopher, an abstract thinker. He has a brain of the first order. He sits motionless, like a spider in the centre of its web, but that web has a thousand radiations, and he knows well every quiver of each of them.

Moriarty exactly matches Holmes's goodness and decency with equal powers of evil. They are mirror images, possessing an essential consonance. (Holmes, too, in 'The Cardboard Box', shares Moriarty's spider-like qualities: 'He loved to lie in the very centre of five millions of people, with his filaments stretching out and running through them, responsive to every little rumour or suspicion of unsolved crime.')

In Moriarty, Conan Doyle seems to have created Holmes's *doppelgänger*: he represents the dark side of the Holmes character; he is the Hyde to Holmes's Jekyll. Since Robert Louis Stevenson (1850–94) was one of Conan Doyle's favourite writers, *Dr Jekyll and Mr Hyde* (1886) may well have been in Conan Doyle's mind when the time came for him to write 'The Final Problem' in 1893. Stevenson and Conan Doyle knew the career of the eminently respectable Edinburgh burglar Deacon William Brodie (executed 1788) and Dickens's unfinished *The Mystery of Edwin Drood* (1870), and they could have drawn on the same sources for the dual-personality theme. If Holmes and Moriarty are seen as two halves of a whole, then there is a certain inevitability in both men plunging to their death in the Reichenbach Falls together.

Whatever Conan Doyle's intentions, 'The Final Problem' is a story that assumes epic proportions in miniature. As a detective story it is weak. There is indeed no detection, for Holmes is occupied purely and simply in the pursuit and destruction of the supreme criminal mastermind. The destruction of Moriarty is to be the crowning point of his career: 'I tell you, Watson, in all seriousness, that if I could beat the man, if I could free society of him, I should feel that my own career had reached its summit.' Viewed as a narrative, however, 'The Final Problem', with all of its action and suspense, is possibly one of the best short stories Conan Doyle ever wrote. The drama of the plot is enhanced by the ever-present malevolence of Professor Moriarty, whose impact on the story is all the more remarkable when it is considered that he never confronts the reader in person. Until 'The Final Problem', the reader, like Watson, has never heard of Professor Moriarty, and in the story itself he is only encountered through Holmes's relating of their meeting in the Baker Street rooms. Watson himself only sees 'a tall man pushing his way furiously through the crowd and waving his hand as if he desired to have the train stopped' and glimpses 'a black figure clearly outlined' on a Swiss hillside. Perhaps this indistinctness supports the view that Holmes and Moriarty were a dual personality and that Conan Doyle, whilst ostensibly intending to bring the Holmes saga to a close, was subconsciously leaving doors ajar through which Holmes could return at a future time. The obvious loose ends in the story—for example, the fact that no bodies were ever recovered and the mention of air-guns early in the story (a reference which was not explained until 'The Empty House' in the *Return*)—suggest that it was Conan Doyle's intention to resurrect Holmes at a later date. He was too shrewd a man, too accomplished an author, not to realize the continuing possibilities of his creation.

The stories of the *Memoirs* are less ostentatious than those of the *Adventures*. In the later series Conan Doyle attempted to show human frailties and, by and large, he succeeded. In the process, he revealed more of the characters and personalities of Holmes and Watson and, as in the case of Henry Wood of 'The Crooked Man', left a memorable impression of the misery which can be caused by the pursuit of selfish ends.

The *Memoirs* reveal a great deal more of the early life of Sherlock Holmes than has previously been shown. 'The "Gloria Scott" ' and 'The Musgrave Ritual' are both stories in which Watson is not involved, except as narrator, as they date back to Holmes's days at college and a few years later. Indeed, the two stories have been the subject of much debate amongst Sherlockians as to exactly which university Holmes attended. Dorothy L. Sayers (*Unpopular Opinions* (1946)) presented an extensive essay on the subject, but her conclusions, like those of succeeding Sherlockian commentators, are only of interest to players of the Sherlockian game. 'The Musgrave Ritual' is notable for the light it throws on Holmes's bohemianism:

> ... he was none the less in his personal habits one of the most untidy men that ever drove a fellow-lodger to distraction ... when I find a man who keeps cigars in the coal-scuttle, his tobacco in the toe end of a Persian slipper, and his unanswered correspondence transfixed by a jack-knife into the very centre of his wooden mantelpiece, then I begin to give myself virtuous airs ... and when Holmes ... would sit in an armchair ... and proceed to adorn the opposite wall with a patriotic V.R. done in bullet-pocks, I felt strongly that neither the atmosphere nor the appearance of our room was improved by it.

'The Greek Interpreter' introduces the only other member of Holmes's family ever encountered in the series: Mycroft Holmes. But that story first tells something of Holmes's own background:

My ancestors were country squires, who appear to have led the same life as is natural to their class. But, none the less, my turn that way is in my veins, and may have come with my grandmother, who was the sister of Vernet, the French artist. Art in the blood is liable to take the strangest forms.

This passage is reminiscent of the way Conan Doyle recounted his own family's history in his autobiography *Memories and Adventures* in 1924. But suggestions that Conan Doyle based Sherlock Holmes on himself can, in the main, be dismissed. As far as the creation of Holmes is concerned, Conan Doyle acknowledged the debt to Dr Joseph Bell, under whom he studied at Edinburgh University. Some of Conan Doyle's sympathies and opinions may well have been transferred to the Holmes character, but the author himself was quick to rebut suggestions that Holmes was too closely aligned to himself. In his famous reply to a letter in verse by Arthur Guiterman (*London Opinion*, vol. 35, December 1912), Conan Doyle wrote:

> ... But is it not on the verge of inanity
> To put down to me my creation's crude vanity?
> He, the created, the puppet of fiction,
> Would not brook rivals nor stand contradiction.
> He, the created, would scoff and would sneer,
> Where I, the creator, would bow and revere.
> So please grip this fact with your cerebral tentacle,
> The doll and its maker are never identical.

The *Memoirs* are Conan Doyle's attempt to make the cold, calculating machine that is Sherlock Holmes appear a little more human. There is a moment in 'The Naval Treaty' when, in the midst of a discussion of the case, Holmes lapses into a totally uncharacteristic mood:

What a lovely thing a rose is ... There is nothing in which deduction is so necessary as in religion ... It can be built up as an exact science by the reasoner. Our highest assurance of the goodness of Providence seems to me to rest in the flowers. All other things, our powers, our desires, our food, are really necessary for our existence in the first instance. But this rose is an extra. Its smell and its colour are an embellishment of life, not a condition of it. It is only goodness which gives extras, and so I say again that we have much to hope from the flowers.

There are echoes of *The Moonstone* in this passage; a recollection of Sergeant Cuff's love of roses:

I haven't much time to be fond of anything . . . But when I *have* a moment's fondness to bestow, most times, Mr Betteredge, the roses get it. . . . If you will look about you . . . you will see that the nature of a man's tastes is, most times, as opposite as possible to the nature of a man's business. Show me any two things more opposite one from the other than a rose and a thief; and I'll correct my tastes accordingly.—

So perhaps Wilkie Collins, though Conan Doyle never directly acknowledged him as an influence, also contributed to the development of Holmes. (Collins was a direct influence on Conan Doyle's early novel *The Mystery of Cloomber* [1888] and, through the development of the plot of that story, on *The Sign of the Four* [1890]).

In the *Memoirs*, Holmes is allowed to make mistakes. An example of him at his blundering worst is 'The Yellow Face', in which Holmes discards his own maxim that 'it is a capital mistake to theorize before one has data'. He assesses the problem outlined to him by Grant Munro and then immediately expounds a theory:

'What do you think of my theory?'
'It is all surmise.'
'But at least it covers all the facts. When new facts come to our knowledge which cannot be covered by it, it will be time enough to reconsider it.'

The ghost of Athelney Jones of *The Sign of the Four* can almost be heard echoing a rebuke: 'Oh, come, now, come! Never be ashamed to own up. . . . Stern facts here—no room for theories.' Holmes is shown to be human after all, aware that even he is not beyond the occasional error; and the close bond of friendship that now exists between Holmes and Watson allows him to admit such infallibility to his colleague: 'Watson . . . if it should ever strike you that I am getting a little over-confident in my powers, or giving less pains to a case than it deserves, kindly whisper "Norbury" in my ear, and I shall be infinitely obliged to you.'

The story of Holmes and Watson is one of developing closeness between two men. This has led to some interpretations of the stories as portraying a homosexual relationship. But these overlook the fact that closeness and love, in a non-sexual form, can demonstrably exist between two men. Homosexual inferences, in fact, say more about modern reticence and the fear of openly displaying affection for another man than about the subtext of the Holmes–Watson relationship. It is obvious, throughout sixty stories, that Conan Doyle never intended to imply a homosexual relationship existing between the two friends, but was simply depicting a close male friendship based on values and modes of expression that are currently unfashionable.

The clear indication of the depth of the relationship existing between Holmes and Watson is beautifully expressed in 'The Final Problem'. In this dramatic narrative one can imagine the horror-stricken Watson racing back to Holmes's assistance at the Reichenbach Falls; the heart-stopping realization that he is too late; the tears he fights back once Holmes's fate is confirmed to him, tears which are all too apparent in the opening and closing lines of the story:

It is with a heavy heart that I take up my pen to write these last few words in which I shall ever record the singular gifts by which my friend Mr Sherlock Holmes was distinguished . . .

. . . him whom I shall ever regard as the best and wisest man whom I have ever known.

Holmes had realized the inevitability of his death throughout the adventure; Watson will always have difficulty in accepting that it was not possible to prevent it.

In the *Memoirs* Conan Doyle attempted to examine in greater detail than before the relationships and motives that lay behind the twelve cases. The stories in this collection are, generally, less light-hearted and more sombre than those of the *Adventures*, and there are fewer elements of comedy. The interaction between men and women is more pronounced in 'The Yellow Face', 'The Musgrave Ritual',

'The Cardboard Box', and 'The Crooked Man' all with plots, or sub-plots, that hinge on a sexual relationship.

In March 1927 the *Strand Magazine* announced a competition for readers to choose the twelve best stories from the forty-four that had already been published in book form. Conan Doyle had already prepared his own list, and the competitor most closely matching his selections would be the winner. Of the twelve stories on Conan Doyle's list, only 'The Final Problem, 'The Musgrave Ritual', and 'The Reigate Squire' appear in the *Memoirs*, and these stories were placed fourth, eleventh, and twelfth respectively. This does not detract from the quality of the stories in the *Memoirs*, nor should it be taken to mean that Conan Doyle did not have a good opinion of the remainder. But some of his comments on the other stories are worth noting. 'Silver Blaze' was excluded because, for all its excellent setting out of clues and deductive reasoning he considered the racing detail to be faulty. For the final two places on the list, he had to make a choice between 'The Bruce-Partington Plans' (*His Last Bow*), 'The Crooked Man', 'The "Gloria Scott" ', 'The Greek Interpreter', 'The Reigate Squire', 'The Musgrave Ritual', and 'The Resident Patient' (*Memoirs*). That four further stories from the *Memoirs*, in addition to the two he selected, were in contention for the final two places on the list is an indication of the quality of this collection. Conan Doyle eventually decided that 'The Musgrave Ritual' should be included as it contained an historical touch giving it a little added distinction and was also a memory from Holmes's early life. 'For the last,' he wrote, 'I might as well draw the name out of a bag, for I see no reason to put one before the other. Whatever their merit—and I make no claim for that—they are all as good as I could make them. On the whole Holmes shows perhaps most ingenuity in "The Reigate Squire", and therefore this shall be the twelfth man in my team.'

T. S. Eliot (1888–1965) seemed to follow Conan Doyle in his favourites when he used several ideas from the stories in the *Memoirs* in his own works. The wording of the Musgrave

Ritual was adapted and transposed for the scene in his play *Murder in the Cathedral* where the second of four Tempters urges Thomas à Becket to submit to King Henry's will, to recover the Chancellorship, and to exercise temporal power for the good of the kingdom. By far the better known borrowings occur in Eliot's poem 'Macavity: The Mystery Cat', which was included in *Old Possum's Book of Practical Cats* (1939). Macavity is obviously Professor Moriarty: 'his head is highly domed . . . He sways his head from side to side, with movements like a snake'; there is a reference to a Treaty going astray, which is, of course, a reference to 'The Naval Treaty'; and the Admiralty's loss of a set of plans is an allusion to 'The Bruce-Partington Plans' (*His Last Bow*). Professor Moriarty is positively identified in the final stanza of the poem:

> And they say that all the Cats whose wicked deeds are
>   widely known
> (I might mention Mungojerrie, I might mention Griddlebone)
> Are nothing more than agents for the Cat who all the time
> Just controls their operations: the Napoleon of Crime!

There seems a deliberate irony in Eliot's elegantly feline adaptation of Moriarty in *Old Possum's Book of Practical Cats* when *Murder in the Cathedral* uses him so effectively in a negative sense. Moriarty's effect on Holmes is to bring him to self-sacrifice not for Glory (as the Fourth Tempter suggests to Becket) or for Art (for which he has so consistently worked in the past), but for Good. Holmes's 'In over a thousand cases I am not aware that I have ever used my powers upon the wrong side' is an understandable self-deception at such a moment (he was very much aware of having done so at the end of 'A Scandal in Bohemia' [*Adventures*], as shown by his refusal of the king's handshake and his bitter reversal of the king's self-differentiation from Irene Adler). But as Holmes himself said of the wretched Silas Brown in 'Silver Blaze', we must have an amnesty in that direction. No doubt Becket needed such a moral amnesty also. Whatever the more dubious details of their careers, they died for the principle of Good.

But if Moriarty is the principle of Evil, his origins are in many respects refreshingly mundane, like so much else in this particular collection lying within the scope of Conan Doyle's observation. The title *Memoirs of Sherlock Holmes* may seem more archaic than any other of the series, since today when we speak of 'memoirs' we mean as a rule such portions of a politician's life as s/he imagines may, if judiciously laundered, bear public scrutiny, whilst some other public figures take themselves seriously enough to use the same term for their autobiographical jottings. Conan Doyle wrote and thought not only in terms of his own day but in that of the old books he hunted and loved so well. His boyhood reading and adult re-reading of Macaulay's *Critical and Historical Essays* placed at his automatic reach forms such as *Memoirs of the Life of Warren Hastings, first Governor-General of Bengal* (1841) by the Revd G. R. Gleig, who had not known his subject and certainly—in Macaulay's opinion—had not understood him. The Holmes *Memoirs* are supposedly from an author who had known Holmes very well, and understood him fairly well, so that an autobiographical element is consciously present; but the focus is on the subject. The title was first added for the English and American book publications; the first American edition included 'The Cardboard Box', dropped in all subsequent editions in conformity with the British publication. In the *Strand*, the future *Memoirs* had simply been entitled 'Adventures of Sherlock Holmes' for the first ('Silver Blaze') and subsequently 'The Adventures of Sherlock Holmes', beginning with the second ('The Cardboard Box'). *The Adventures of Sherlock Holmes* never had a definite article in its *Strand* form, but received it on book publication six weeks before 'Silver Blaze' commenced in the *Strand*; its American edition also omitted the article, and the firm insertion of that 'The', first in book, then in magazine, was the very first sign in print that Conan Doyle was putting a definite halt on what the *Strand* editor had previously presented as a commodity with unlimited supply. The first American *Memoirs* (again differing from the British with no restrictive 'The') announced its origin by the running title 'The Adventures of Sherlock Holmes', the correct 'Memoirs of Sherlock Holmes' only appearing as running title in the 'New and Revised Edition', which like many another product blazoned as 'New' gave the customer less for the same money (in this case $1.50). The result is that while the purchaser of a British first edition of the *Memoirs* has obtained an attractive item, the possessor of a first American edition may be considered a 'fortunate owner' as fully as Colonel Ross of 'Silver Blaze'.

Yet *The Memoirs of Sherlock Holmes*, however belatedly entitled, was in some sense Conan Doyle's own memoirs. He was moving towards an autobiographical impulse, realized in 1895 with *The Stark Munro Letters*. The original *Adventures* drew on his experiences also, of course, but show a tense grasping at this new London where he had just settled and where he so impudently had settled his Holmes and Watson several years before his own descent on it. The *Memoirs* are far less London-centred, moving far afield not only in space but in time. But the fields of personal experience on which Conan Doyle drew were cumulative. There are two sea narratives, one the sailor Jim Browner's statement of his shortest and most terrible voyage in 'The Cardboard Box', and the other the convict James Armitage's 'Some particulars of the voyage of the bark *Gloria Scott*'. Conan Doyle had already shown his prowess with tales of blue water—witness his classical ghost story 'The Captain of the *Pole Star*', which combined Greenland whaling with Emily Brontë, and 'J. Habakuk Jephson's Statement', combining his voyage with the *Mayumba* in African waters with inspiration from H. H. Garnet to produce his fictional solution to the great sea mystery of the intact but unpeopled *Mary Celeste*. The vivid quality of his writing, which has made Holmes and Watson more real to Sherlockians than are most historical person-ages, induced conviction that the Statement was the work of a real Habakuk Jephson; and the sea stories in the *Memoirs* put us inescapably in the element, whether by means of the early nineteenth-century convict transportation or the late nineteenth-century seaside pleasure-boats. (One wonders how many people wanted to take trips on boats from the Parade after reading 'The Cardboard Box'.) As a reader, his work was enhanced by his devotion to Herman Melville and Clark Russell, not to speak of Stevenson (Prendergast and his genial chaplain seem ideal companions for Job Anderson and Long John Silver, even if Prendergast is a little fast with his naval terms for a landsman). Conan Doyle as a doctor in Southsea was in a good situation to refurbish his sea memories. Old soldiers, too, were natural patients to supply at least the medical basis of the anguish of Henry Wood in 'The Crooked Man'.

'The Stockbroker's Clerk' is in some ways a disappointing story: Holmes does little save give what proves bad advice to go to Birmingham instead of making an investigation in London to see who has turned up at Mawson & Williams. This may have been the reason for writing it: Conan Doyle wanted to pay some homage to Birmingham, where he had been so happy as an assistant and extra-mural student in Pharmacology to Dr Reginald Ratcliff Hoare in Aston on various occasions between 1879 and 1882. It is clear he was also proud, as a new Londoner, of his grasp of Cockney slang, in which Hall Pycroft displays as much expertise as in his stock exchange data; there is precision in the selection of terms, and their distinction from underworld slang of which Conan Doyle, whether from patients or elsewhere, also showed a fair acquaintance. A doctor's observation is present in the emphasis on family affection of Jews, also very movingly asserted in a greater story of late Victorian finance, Anthony Trollope's *The Way We Live Now*. The heartbreak of Pinner alias Beddington is allowed to close the story, as does that of the widowed Godfrey Staunton in 'The Missing Three-Quarter' (*Return*): the absence of the usual coda in both cases is clearly deliberate. The racial sensitivity may, like that of 'The Yellow Face', owe something to what Conan Doyle learned from Henry Highland Garnet; but it makes total sense to have Pinner's Jewish origins announced with Hall Pycroft's insensitive vulgarity. The deployment of Pycroft's Cockney speech gives him the unusual status of being more audible than the normal Sherlock Holmes client, and however different from his creator in manner, he shared his bitter hunger for employment. Otherwise that hunger is a feature more of the *Adventures*, save that Holmes gives us a taste of his years of underemployment in 'The Musgrave Ritual', and 'The Resident Patient' may recall a delicious dream of how a near-starving doctor might hope for some successful bank-robber to set him up in practice. The foredoomed trial of Blessington by his old acquaintance, who declined to be forgot, sounds like a chilling memory from Conan Doyle's Irish relatives about Fenian treatment of informers and discredited leaders. It takes up a possible irritant for Conan Doyle in other detective fiction where the guilty party is blasted by the avenger into Kingdom Come without a moment's suffering beyond the instant of death. The thoughts of the gagged and pathologically terrorized Blessington while facing his indictment, judges, and hangmen, are not pleasant to contemplate. One of the deeper effects of the stories is the scientific ticking-off of data by Holmes, the implications of which are as horrific as a list of symptoms may be for a fatal disease:

It must have lasted for some time, for it was then that these cigars were smoked. The older man sat in that wicker-chair; it was he who used the cigar-holder. The younger man sat over yonder; he knocked his ash off against the chest of drawers. The third fellow paced up and down. Blessington, I think, sat upright in the bed, but of that I cannot be absolutely certain.

'The Musgrave Ritual' is evidently a memory of the wealthy Catholic Lancashire squires of Stonyhurst College, encountered by their impoverished Edinburgh schoolmate son of a drunken local government officer; it is perceptive, possibly retrospectively, in recognizing the shyness that often hid behind social pretension. But the brother of governess sisters at the mercy of an omnipotent employer knew the danger of being turned off without a testimonial at some scarcely conceivable cause for offence, regardless of previous service. It is a chilling reminder of realities of the time.

Stonyhurst gossip about estate litigation, from which the school itself suffered in ACD's time, seems an origin of the Acton–Cunningham suit in 'The Reigate Squire', and Alec Cunningham's methods of dealing with the unwanted evidence seem sufficiently like a solution bruited in schoolboys' chatter. No doubt Holmes's malingering might also have a school origin, but a doctor's practice gave adequate grounding in that. The device by which Alec implicates his father (a delicate illustration of their mutual confidence), the alternate writing of words in the death-bait message, has also a touch of schoolboy insurance: the use to which it is put, in showing the contrast of handwritings in the seven surviving words in the fragment, and then their similarity, is as impressive a piece of jam-spreading on both sides of the bread as one could ask. Stonyhurst tradition also credits the Prefect of Discipline, the Revd Thomas Kay, SJ, in whose hands lay correction of schoolboy offenders, with the manner of Moriarty at his meeting with Holmes. (The 'loaded firearm in the pocket of one's dressing-gown' might have been a 'lighted pipe', the use of which was a frequent cause of Kay's unsympathetic interest in the boy Doyle.) On the other hand, the presence of two Moriartys at Stonyhurst in Conan Doyle's time seems doubly suggestive: Michael Moriarty won a mathematical prize (possibly for work on the binomial theorem) when in his first year of secondary school against entrants from all other classes; John Francis Moriarty, afterwards Lord Justice of Ireland, was remembered by contemporaries for his devious architecture of self-serving intrigue and his serpentine oscillation in conversational exchanges. Mycroft Holmes in 'The Greek Interpreter', on the other hand, seems like a memory of research specialists in Edinburgh University medical school; the portrait is satirical, as may be judged by the practical intervention that causes the death of Paul Kratides and very nearly that of his client, Melas. Were it not for his brother's intervention, Mycroft would chase off after gossip about Sophie regardless of the others' mortal peril. There is ample academic parallel for those priorities also.

'The Greek Interpreter' opens up another question, though how far it is memoir for Conan Doyle we may never know. Catherine Belsey, in her instructive *Critical Practice* (1980), expresses it strikingly:

These stories, whose overt project is total explicitness, total verisimilitude in the interests of a plea for scientificity, are haunted by shadowy, mysterious and often silent women. Their silence repeatedly conceals their sexuality, investing it with a dark and magical quality which is beyond the reach of scientific knowledge. In 'The Greek Interpreter' Sophie Kratides has run away with a man. Though she is the pivot of the plot she appears only briefly: 'I could not see her clearly enough to know more than that she was tall and graceful, with black hair, and clad in some sort of loose white gown.' Connotatively the white gown marks her as still virginal and her flight as the result of romance rather than desire. At the same time the dim light surrounds her with shadow, the unknown. 'The Crooked Man' concerns Mrs Barclay, whose husband is found dead on the day of her meeting with her lover of many years before. Mrs Barclay is now insensible, 'temporarily insane' since the night of the murder and therefore unable to speak ... The classic realist text had not yet developed a way of signifying women's sexuality except in a metaphoric or symbolic mode whose presence disrupts the realist surface ... What is more significant, however, is that the presentation of so many women in the Sherlock Holmes stories as shadowy, mysterious and magical figures precisely contradicts the project of explicitness, transgresses the values of the texts, and in doing so throws into relief the poverty of the contemporary concept of science.

But this was Conan Doyle's intention: his whole life was an expression of the tension of science and spirit in which neither should claim the victory. His poem 'The Inner Room' proclaims as much. Indeed in 'The Naval Treaty' Holmes not only makes his great credo of the rose but, contrary to all his principles and prejudices, stakes everything on Annie Harrison's love for her fiancé proving stronger than any pressure from her brother. It was probably as a reflective brother that he wrote of the strength of such women in a world of weak men like 'Tadpole' Phelps and selfish men like Joseph Harrison. From the fragments of evidence that remain to us, it is possible to trace some hint of a resemblance here to his eldest sister Annette, a governess in Portugal who brought the next sisters to join her there and played a brave part through the breakup of the family, only to die in 1889.

The final sense of the title lies not with Conan Doyle but Sherlock Holmes. Holmes really does produce memoirs within these pages, and a distinguished narrator he proves. The idea of Holmes as storyteller is foreign to the display of his talents hitherto, however eloquently he may recount his train of reasoning. In the *Adventures* and the previous two novels, Watson keeps pace with Holmes too well to give us any description of events of any length, although here 'The Naval Treaty' and 'The Final Problem' both afford excellent examples of his narrative skills. But his power of telling a full story within the Watsonian framework (as distinct from the two Holmes narratives of the *Case-Book*) is shown only in 'The "Gloria Scott" ' and 'The Musgrave Ritual', and the little self-revelations, especially on the meaning of Vocation, that accompany them, are very different from his more public or professional demonstrations:

And that recommendation, with the exaggerated estimate of my abilities with which he prefaced it, was, if you will believe me, Watson, the very first thing which ever made me feel that a profession might be made out of what had up to that time been the merest hobby.'

or

'But I understand, Holmes, that you are turning to practical ends those powers with which you used to amaze us?'

   'Yes', said I, 'I have taken to living by my wits.'

CHRISTOPHER RODEN

NOTE ON THE TEXT

Except for 'The Cardboard Box' and 'The Resident Patient', which are derived from the *Strand Magazine* (see Explanatory Notes), the text for this edition is based on the first book edition of the *Memoirs* (1893), published by George Newnes, collated with the *Strand* and with American printings. Where reason has dictated the adoption of an alternative reading from Newnes 1893, it has been noted, save for punctuation.

SELECT BIBLIOGRAPHY

I. A. CONAN DOYLE: PRINCIPAL WORKS

(a) *Fiction*

A Study in Scarlet (Ward, Lock, & Co., 1888)
The Mystery of Cloomber (Ward & Downey, 1888)
Micah Clarke (Longmans, Green, & Co., 1889)
The Captain of the Pole-Star and Other Tales (Longmans, Green, & Co., 1890)
The Sign of the Four (Spencer Blackett, 1890)
The Firm of Girdlestone (Chatto & Windus, 1890)
The White Company (Smith, Elder, & Co., 1891)
The Adventures of Sherlock Holmes (George Newnes, 1892)
The Great Shadow (Arrowsmith, 1892)
The Refugees (Longmans, Green, & Co., 1893)
The Memoirs of Sherlock Holmes (George Newnes, 1893)
Round the Red Lamp (Methuen & Co., 1894)
The Stark Munro Letters (Longmans, Green, & Co., 1895)
The Exploits of Brigadier Gerard (George Newnes, 1896)
Rodney Stone (Smith, Elder, & Co., 1896)
Uncle Bernac (Smith, Elder, & Co., 1897)
The Tragedy of the Korosko (Smith, Elder, & Co., 1898)
A Duet With an Occasional Chorus (Grant Richards, 1899)
The Green Flag and Other Stories of War and Sport (Smith, Elder, & Co., 1900)
The Hound of the Baskervilles (George Newnes, 1902)
Adventures of Gerard (George Newnes, 1903)
The Return of Sherlock Holmes (George Newnes, 1905)
Sir Nigel (Smith, Elder, & Co., 1906)
Round the Fire Stories (Smith, Elder, & Co., 1908)
The Last Galley (Smith, Elder, & Co., 1911)
The Lost World (Hodder & Stoughton, 1912)
The Poison Belt (Hodder & Stoughton, 1913)
The Valley of Fear (Smith, Elder, & Co., 1915)
His Last Bow (John Murray, 1917)
Danger! and Other Stories (John Murray, 1918)
The Land of Mist (Hutchinson & Co., 1926)
The Case-Book of Sherlock Holmes (John Murray, 1927)
The Maracot Deep and Other Stories (John Murray, 1929)

The Complete Sherlock Holmes Short Stories (John Murray, 1928)
The Conan Doyle Stories (John Murray, 1929)
The Complete Sherlock Holmes Long Stories (John Murray, 1929)

(b) *Non-fiction*

The Great Boer War (Smith, Elder, & Co., 1900)
The Story of Mr George Edalji (T. Harrison Roberts, 1907)
Through the Magic Door (Smith, Elder, & Co., 1907)
The Crime of the Congo (Hutchinson & Co., 1909)
The Case of Oscar Slater (Hodder & Stoughton, 1912)
The German War (Hodder & Stoughton, 1914)
The British Campaign in France and Flanders (Hodder & Stoughton, 6 vols., 1916–20)
The Poems of Arthur Conan Doyle (John Murray, 1922)
Memories and Adventures (Hodder & Stoughton, 1924; revised ed., 1930)
The History of Spiritualism (Cassell & Co., 1926)

2. MISCELLANEOUS

A Bibliography of A. Conan Doyle (Soho Bibliographies 23: Oxford, 1983) by Richard Lancelyn Green and John Michael Gibson, with a foreword by Graham Greene, is the standard—and indispensable—source of bibliographical information, and of much else besides. Green and Gibson have also assembled and introduced *The Unknown Conan Doyle*, comprising *Uncollected Stories* (those never previously published in book form); *Essays in Photography* (documenting a little-known enthusiasm of Conan Doyle's during his time as a student and young doctor), both published in 1982; and *Letters to the Press* (1986). Alone, Richard Lancelyn Green has compiled (1) *The Uncollected Sherlock Holmes* (1983), an impressive assemblage of Holmesiana, containing almost all Conan Doyle's writing about his creation (other than the stories themselves) together with related material by Joseph Bell, J. M. Barrie, and Beverley Nichols; (2) *The Further Adventures of Sherlock Holmes* (1985), a selection of eleven apocryphal Holmes adventures by various authors, all diplomatically introduced; (3) *The Sherlock Holmes Letters* (1986), a collection of noteworthy public correspondence on Holmes and Holmesiana and far more valuable than its title suggests; and (4) *Letters to Sherlock Holmes* (1984), a powerful testimony to the power of the Holmes stories.

Though much of Conan Doyle's work is now readily available there are still gaps. Some of his very earliest fiction now only survives in rare piracies (apart, that is, from the magazines in which they were first published), including items of intrinsic genre interest such as 'The Gully of Bluemansdyke' (1881) and its sequel 'My Friend the Murderer' (1882), which both turn on the theme of the murderer-informer (handled very differently—and far better—in the Holmes story of 'The Resident Patient' (*Memoirs*)): both of these were used as book-titles for the same pirate collection first issued as *Mysteries and Adventures* (1889). Other stories achieved book publication only after severe pruning—for example, 'The Surgeon of Gaster Fell', reprinted in *Danger!* many years after magazine publication (1890). Some items given initial book publication were not included in the collected edition of *The Conan Doyle Stories*. Particularly deplorable losses were 'John Barrington Cowles' (1884: included subsequently in *Edinburgh Stories of Arthur Conan Doyle* (1981)), 'A Foreign Office Romance' (1894), 'The Club-Footed Grocer' (1898), 'A Shadow Before' (1898), and 'Danger!' (1914). Three of these may have been post-war casualties, as seeming to deal too lightheartedly with the outbreak of other wars; 'John Barrington Cowles' may have been dismissed as juvenile work; but why Conan Doyle discarded a story as good as 'The Club-Footed Grocer' would baffle even Holmes.

At the other end of his life, Conan Doyle's tidying impaired the survival of his most recent work, some of which well merited lasting recognition. *The Maracot Deep and Other Stories* appeared in 1929, a little over a month after *The Conan Doyle Stories*; 'Maracot' itself found a separate paperback life as a short novel; the two Professor Challenger stories, 'The Disintegration Machine' and 'When the World Screamed', were naturally included in John Murray's *The Professor Challenger Stories* (1952); but the fourth item, 'The Story of Spedegue's Dropper', passed beyond the ken of most of Conan Doyle's readers. These three stories show the author, in his seventieth year, still at the height of his powers.

In 1980 Gaslight Publications, of Bloomington, Ind., reprinted *The Mystery of Cloomber*, *The Firm of Girdlestone*, *The Doings of Raffles Haw* (1892), *Beyond the City* (1893), *The Parasite* (1894; also reprinted in *Edinburgh Stories of Arthur Conan Doyle*), *The Stark Munro Letters*, *The Tragedy of the Korosko*, and *A Duet*. *Memories and Adventures*, Conan Doyle's enthralling but impressionistic recollections, are best read in the revised (1930) edition. *Through the Magic Door* remains the best introduction to the literary mind of Conan Doyle, whilst some of his volumes on Spiritualism have autobiographical material of literary significance.

ACD: The Journal of the Arthur Conan Doyle Society (ed. Christopher Roden, David Stuart Davies [to 1991], and Barbara Roden [from 1992]), together with its newsletter, *The Parish Magazine*, is a useful source of critical and biographical material on Conan Doyle. The enormous body of 'Sherlockiana' is best pursued in *The Baker Street Journal*, published by Fordham University Press, or in the *Sherlock Holmes Journal* (Sherlock Holmes Society of London), itemized up to 1974 in the colossal *World Bibliography of Sherlock Holmes and Doctor Watson* (1974) by Ronald Burt De Waal (see also De Waal, *The International Sherlock Holmes* (1980)) and digested in *The Annotated Sherlock Holmes* (2 vols., 1968) by William S. Baring-Gould, whose industry has been invaluable for the Oxford Sherlock Holmes editors. Jack Tracy, *The Encyclopaedia Sherlockiana* (1979) is a very helpful compilation of relevant data. Those who can nerve themselves to consult it despite its title will benefit greatly from Christopher Redmond, *In Bed With Sherlock Holmes* (1984). The classic 'Sherlockian' work is Ronald A. Knox, 'Studies in the Literature of Sherlock Holmes', first published in *The Blue Book* (July 1912) and reprinted in his *Essays in Satire* (1928).

The serious student of Conan Doyle may perhaps deplore the vast extent of 'Sherlockian' literature, even though the size of this output is testimony in itself to the scale and nature of Conan Doyle's achievement. But there is undoubtedly some wheat amongst the chaff. At the head stands Dorothy L. Sayers, *Unpopular Opinions* (1946); also of some interest are T. S. Blakeney, *Sherlock Holmes: Fact or Fiction* (1932), H. W. Bell, *Sherlock Holmes and Dr Watson* (1932), Vincent Starrett, *The Private Life of Sherlock Holmes* (1934), Gavin Brend, *My Dear Holmes* (1951), S. C. Roberts, *Holmes and Watson* (1953) and Roberts's introduction to *Sherlock Holmes: Selected Stories* (Oxford: The World's Classics, 1951), James E. Holroyd, *Baker Street Byways* (1959), Ian McQueen, *Sherlock Holmes Detected* (1974), and Trevor H. Hall, *Sherlock Holmes and his Creator* (1978). One Sherlockian item certainly falls into the category of the genuinely essential: D. Martin Dakin, *A Sherlock Holmes Commentary* (1972), to which all the editors of the present series are indebted.

Michael Pointer, *The Public Life of Sherlock Holmes* (1975) contains invaluable information concerning dramatizations of the Sherlock Holmes stories for radio, stage, and the cinema; of complementary interest are Chris Steinbrunner and Norman Michaels, *The Films of Sherlock Holmes* (1978) and David Stuart Davies, *Holmes of the Movies* (1976), whilst Philip Weller with Christopher Roden, *The Life and Times of Sherlock Holmes* (1992) summarizes a great deal of useful information concerning Conan Doyle's life and Holmes's cases, and in addition is delightfully illustrated. The more concrete products of the Holmes industry are dealt with in Charles Hall, *The Sherlock Holmes Collection* (1987). For a useful retrospective view, Allen Eyles, *Sherlock Holmes: A Centenary Celebration* (1986) rises to the occasion. Both useful and engaging are Peter Haining, *The Sherlock Holmes Scrapbook* (1973) and Charles Viney, *Sherlock Holmes in London* (1989).

Of the many anthologies of Holmesiana, P. A. Shreffler (ed.), *The Baker Street Reader* (1984) is exceptionally useful. D. A. Redmond, *Sherlock Holmes: A Study in Sources* (1982) is similarly indispensable. Michael Hardwick, *The Complete Guide to Sherlock Holmes* (1986) is both reliable and entertaining; Michael Harrison, *In the Footsteps of Sherlock Holmes* (1958) is occasionally helpful.

For more general studies of the detective story, the standard history is Julian Symons, *Bloody Murder* (1972, 1985, 1992). Necessary but a great deal less satisfactory is Howard Haycraft, *Murder for Pleasure* (1942); of more value is Haycraft's critical anthology *The Art of the Mystery Story* (1946), which contains many choice period items. Both R. F. Stewart, *. . . And Always a Detective* (1980) and Colin Watson, *Snobbery with Violence* (1971) are occasionally useful. Dorothy Sayers's pioneering introduction to *Great Short Stories of Detection, Mystery and Horror* (First Series, 1928), despite some inspired howlers, is essential reading; Raymond Chandler's riposte, 'The Simple Art of Murder' (1944), is reprinted in Haycraft, *The Art of the Mystery Story* (see above). Less well known than Sayers's essay but with an equal claim to poineer status is E. M. Wrong's introduction to *Crime and Detection*, First Series (Oxford: The World's Classics, 1926). See also Michael Cox (ed.), *Victorian Tales of Mystery and Detection: An Oxford Anthology* (1992).

Amongst biographical studies of Conan Doyle one of the most distinguished is Jon L. Lellenberg's survey, *The Quest for Sir Arthur Conan Doyle* (1987), with a Foreword by Dame Jean Conan Doyle (much the best piece of writing on ACD by any member of his family). The four earliest biographers—the Revd John Lamond (1931), Hesketh Pearson (1943), John Dickson Carr (1949), and Pierre Nordon (1964)—all had access to the family archives, subsequently closed to researchers following a lawsuit; hence all four biographies contain valuable documentary material, though Nordon handles the evidence best (the French text is fuller than the English version, published in 1966). Of the others, Lamond seems only to have made little use of the material available to him;

Pearson is irreverent and wildly careless with dates; Dickson Carr has a strong fictionalizing element. Both he and Nordon paid a price for their access to the Conan Doyle papers by deferring to the far from impartial editorial demands of Adrian Conan Doyle; Nordon nevertheless remains the best available biography. The best short sketch is Julian Symons, *Conan Doyle* (1979) (and for the late Victorian milieu of the Holmes cycle some of Symons's own fiction, such as *The Blackheath Poisonings* and *The Detling Secret*, can be thoroughly recommended). Harold Orel (ed.), *Critical Essays on Sir Arthur Conan Doyle* (1992) is a good and varied collection, whilst Robin Winks, *The Historian as Detective* (1969) contains many insights and examples applicable to the Holmes corpus; Winks's *Detective Fiction: A Collection of Critical Essays* (1980) is an admirable working handbook, with a useful critical bibliography. Edmund Wilson's famous essay 'Mr Holmes, they were the footprints of a gigantic hound' (1944) may be found in his *Classics and Commercials: A Literary Chronicle of the Forties* (1950).

Specialized biographical areas are covered in Owen Dudley Edwards, *The Quest for Sherlock Holmes: A Biographical Study of Arthur Conan Doyle* (1982) and in Geoffrey Stavert, *A Study in Southsea: The Unrevealed Life of Dr Arthur Conan Doyle* (1987), which respectively assess the significance of the years up to 1882, and from 1882 to 1890. Alvin E. Rodin and Jack D. Key provide a thorough study of Conan Doyle's medical career and its literary implications in *Medical Casebook of Dr Arthur Conan Doyle* (1984). Peter Costello, in *The Real World of Sherlock Holmes: The True Crimes Investigated by Arthur Conan Doyle* (1991) claims too much, but it is useful to be reminded of events that came within Conan Doyle's orbit, even if they are sometimes tangential or even irrelevant. Christopher Redmond, *Welcome to America, Mr Sherlock Holmes* (1987) is a thorough account of Conan Doyle's tour of North America in 1894.

Other than Baring-Gould (see above), the only serious attempt to annotate the nine volumes of the Holmes cycle has been in the Longman Heritage of Literature series (1979–80), to which the present editors are also indebted. Of introductions to individual texts, H. R. F. Keating's to the *Adventures* and *The Hound of the Baskervilles* (published in one volume under the dubious title *The Best of Sherlock Holmes* (1992)) is worthy of particular mention.

A CHRONOLOGY OF ARTHUR CONAN DOYLE

1855 Charles Altamont Doyle, youngest son of the political cartoonist John Doyle ('HB'), and Mary Foley, his Irish landlady's daughter, marry in Edinburgh on 31 July.

1859 Arthur Ignatius Conan Doyle, third child and elder son of ten siblings, born at 11 Picardy Place, Edinburgh, on 22 May and baptized into the Roman Catholic religion of his parents.

1868–75 ACD commences two years' education under the Jesuits at Hodder, followed by five years at its senior sister college, Stonyhurst, both in the Ribble Valley, Lancashire; becomes a popular storyteller amongst his fellow pupils, writes verses, edits a school paper, and makes one close friend, James Ryan of Glasgow and Ceylon. Doyle family resides at 3 Sciennes Hill Place, Edinburgh.

1875–6 ACD passes London Matriculation Examination at Stonyhurst and studies for a year in the Jesuit college at Feldkirch, Austria.

1876–7 ACD becomes a student of medicine at Edinburgh University on the advice of Bryan Charles Waller, now lodging with the Doyle family at 2 Argyle Park Terrace.

1877–80 Waller leases 23 George Square, Edinburgh as a 'consulting pathologist', with all the Doyles as residents. ACD continues medical studies, becoming surgeon's clerk to Joseph Bell at Edinburgh; also takes temporary medical assistantships at Sheffield, Ruyton (Salop), and Birmingham, the last leading to a close friendship with his employer's family, the Hoares. First story published, 'The Mystery of Sasassa Valley', in *Chambers's Journal* (6 Sept. 1879); first non-fiction published—'Gelseminum as a Poison', *British Medical Journal* (20 Sept. 1879). Some time previously ACD sends 'The Haunted Grange of Goresthorpe' to *Blackwood's Edinburgh Magazine*, but it is filed and forgotten.

1880 (Feb.–Sept.) ACD serves as surgeon on the Greenland whaler *Hope* of Peterhead.

1881   ACD graduates MB, CM (Edin.); Waller and the Doyles living at 15 Lonsdale Terrace, Edinburgh.

1881–2 (Oct.–Jan.) ACD serves as surgeon on the steamer *Mayumba* to West Africa, spending three days with US Minister to Liberia, Henry Highland Garnet, black abolitionist leader, then dying. (July–Aug.) Visits Foley relatives in Lismore, Co. Waterford.

1882   Ill-fated partnership with George Turnavine Budd in Plymouth. ACD moves to Southsea, Portsmouth, in June. ACD published in *London Society*, *All the Year Round*, *Lancet*, and *British Journal of Photography*. Over the next eight years ACD becomes an increasingly successful general practitioner at Southsea.

1882–3 Breakup of the Doyle family in Edinburgh. Charles Altamont Doyle henceforth confined because of alcoholism and epilepsy. Mary Foley Doyle resident in Masongill Cottage on the Waller estate at Masongill, Yorkshire. Innes Doyle (b. 1873) resident with ACD as schoolboy and surgery page from Sept. 1882.

1883   'The Captain of the *Pole-Star*' published (*Temple Bar*, Jan.), as well as a steady stream of minor pieces. Works on *The Mystery of Cloomber*.

1884   ACD publishes 'J. Habakuk Jephson's Statement' (*Cornhill Magazine*, Jan.), 'The Heiress of Glenmahowley' (*Temple Bar*, Jan.), 'The Cabman's Story' (*Cassell's Saturday Journal*, May); working on *The Firm of Girdlestone*.

1885   Publishes 'The Man from Archangel' (*London Society*, Jan.). Jack Hawkins, briefly a resident patient with ACD, dies of cerebral meningitis. Louisa Hawkins, his sister, marries ACD. (Aug.) Travels in Ireland for honeymoon. Awarded Edinburgh MD.

1886   Writing *A Study in Scarlet*.

1887   *A Study in Scarlet* published in *Beeton's Christmas Annual*.

1888   (July) First book edition of *A Study in Scarlet* published by Ward, Lock; (Dec.) *The Mystery of Cloomber* published.

1889   (Feb.) *Micah Clarke* (ACD's novel of the Monmouth Rebellion of 1685) published. Mary Louise Conan Doyle, ACD's eldest child, born. Unauthorized publication of *Mysteries and Adventures* (published later as *The*

Gully of Bluemansdyke and *My Friend the Murderer*). *The Sign of the Four* and Oscar Wilde's *The Picture of Dorian Gray* commissioned by Lippincott's.

1890   (Jan.) 'Mr [R. L.] Stevenson's Methods in Fiction' published in the *National Review*. (Feb.) *The Sign of the Four* published in *Lippincott's Monthly Magazine*; (Mar.) First authorized short-story collection, *The Captain of the Pole-Star and Other Tales*, published; (Apr.) *The Firm of Girdlestone* published; (Oct.) First book edition of the *Sign* published by Spencer Blackett.

1891   ACD sets up as an eye specialist in 2 Upper Wimpole Street, off Harley Street, while living at Montague Place. Moves to South Norwood. (July–Dec.) The first six 'Adventures of Sherlock Holmes' published in George Newnes's *Strand Magazine*. (Oct.) *The White Company* published; *Beyond the City* first published in *Good Cheer*, the special Christmas number of *Good Words*.

1892   (Jan.–June) Six more Holmes stories published in the *Strand*, with another in Dec. (Mar.) *The Doings of Raffles Haw* published (first serialized in Alfred Harmsworth's penny paper *Answers*, Dec. 1891–Feb. 1892). (14 Oct.) *The Adventures of Sherlock Holmes* published by Newnes. (31 Oct.) Waterloo story *The Great Shadow* published. Alleyne Kingsley Conan Doyle born. Newnes republishes the *Sign*.

1893   'Adventures of Sherlock Holmes' (second series) continues in the *Strand*, to be published by Newnes as *The Memoirs of Sherlock Holmes* (Dec.), minus 'The Cardboard Box'. Holmes apparently killed in 'The Final Problem' (Dec.) to free ACD for 'more serious literary work'. (May) *The Refugees* published. *Jane Annie: or, the Good Conduct Prize* (musical comedy co-written with J. M. Barrie) fails at the Savoy Theatre. (10 Oct.) Charles Altamont Doyle dies.

1894   (Oct.) *Round the Red Lamp*, a collection of medical short stories, published, several for the first time. *The Stark Munro Letters*, a fictionalized autobiography, begun, to be concluded the following year. ACD on US lecture tour with Innes Doyle. (Dec.) *The Parasite* published; 'The Medal of Brigadier Gerard' published in the *Strand*.

1895   'The Exploits of Brigadier Gerard' published in the *Strand*.

1896   (Feb.) *The Exploits of Brigadier Gerard* published by Newnes. ACD settles at Hindhead, Surrey, to minimize effects of his wife's tuberculosis. (Nov.) *Rodney Stone*, a pre-Regency mystery, published. Self-pastiche, 'The Field Bazaar', appears in the Edinburgh University *Student* (20 Nov.).

1897   (May) Napoleonic novel *Uncle Bernac* published; three 'Captain Sharkey' pirate stories published in *Pearson's Magazine* (Jan., Mar., May). Home at Undershaw, Hindhead.

1898   (Feb.) *The Tragedy of the Korosko* published. (June) Publishes *Songs of Action*, a verse collection. (June–Dec.) Begins to publish 'Round the Fire Stories' in the *Strand*—'The Beetle Hunter', 'The Man with the Watches', 'The Lost Special', 'The Sealed Room', 'The Black Doctor', 'The Club-Footed Grocer', and 'The Brazilian Cat'. Ernest William Hornung (ACD's brother-in-law) creates A. J. Raffles and in 1899 dedicates the first stories to ACD.

1899   (Jan.–May) Concludes 'Round the Fire' series in the *Strand* with 'The Japanned Box', 'The Jew's Breast-Plate', 'B. 24', 'The Latin Tutor', and 'The Brown Hand'. (Mar.) Publishes *A Duet with an Occasional Chorus*, a version of his own romance. (Oct.–Dec.) 'The Croxley Master', a boxing story, published in the *Strand*. William Gillette begins 33 years starring in *Sherlock Holmes*, a play by Gillette and ACD.

1900   Accompanies volunteer-staffed Langman hospital as unofficial supervisor to support British forces in the Boer War. (Mar.) Publishes short-story collection, *The Green Flag and Other Stories of War and Sport*. (Oct.) *The Great Boer War* published. Unsuccessful Liberal Unionist parliamentary candidate for Edinburgh Central.

1901   (Aug.) 'The Hound of the Baskervilles' begins serialization in the *Strand*, subtitled 'Another Adventure of Sherlock Holmes'.

1902   (Jan.) *The War in South Africa: Its Cause and Conduct* published. 'Sherlockian' higher criticism begun by Frank Sidgwick in the *Cambridge Review* (23 Jan.). (Mar.) *The Hound of the Baskervilles* published by Newnes. ACD accepts knighthood with reluctance.

1903   (Sept.) *Adventures of Gerard* published by Newnes (previously serialized in the *Strand*). (Oct.) 'The Return of

Sherlock Holmes' begins in the *Strand*. Author's Edition of ACD's major works published in twelve volumes by Smith, Elder and thirteen by D. Appleton & Co. of New York, with prefaces by ACD; many titles omitted.

1904   'Return of Sherlock Holmes' continues in the *Strand*; series designed to conclude with 'The Abbey Grange' (Sept.), but ACD develops earlier allusions and produces 'The Second Stain' (Dec.).

1905   (Mar.) *The Return of Sherlock Holmes* published by Newnes. (Dec.) Serialization of 'Sir Nigel' begun in the *Strand* (concluded Dec. 1906).

1906   (Nov.) Book publication of *Sir Nigel*. ACD defeated as Unionist candidate for Hawick District in general election. (4 July) Death of Louisa ('Touie'), Lady Conan Doyle. ACD deeply affected.

1907   ACD clears the name of George Edalji (convicted in 1903 of cattle-maiming). (18 Sept.) Marries Jean Leckie. (Nov.) Publishes *Through the Magic Door*, a celebration of his literary mentors (earlier version serialized in *Great Thoughts*, 1894).

1908   Moves to Windlesham, Crowborough, Sussex. (Jan.) Death of Sidney Paget. (Sept.) *Round the Fire Stories* published, including some not in earlier *Strand* series. (Sept.–Oct.) 'The Singular Experience of Mr John Scott Eccles' (later retitled as 'The Adventure of Wisteria Lodge') begins occasional series of Holmes stories in the *Strand*.

1909   ACD becomes President of the Divorce Law Reform Union (until 1919). Denis Percy Stewart Conan Doyle born. Takes up agitation against Belgian oppression in the Congo.

1910   (Sept.) 'The Marriage of the Brigadier', the last Gerard story, published in the *Strand*, and (Dec.) the Holmes story of 'The Devil's Foot'. ACD takes six-month lease on Adelphi Theatre; the play *The Speckled Band* opens there, eventually running to 346 performances. Adrian Malcolm Conan Doyle born.

1911   (Apr.) *The Last Galley* (short stories, mostly historical) published. Two more Holmes stories appear in the *Strand*: 'The Red Circle' (Mar., Apr.) and 'The Disappearance of Lady Frances Carfax' (Dec.). ACD declares for Irish Home Rule, under the influence of Sir Roger Casement.

1912   (Apr.–Nov.) The first Professor Challenger story, *The Lost World*, published in the *Strand*, book publication in Oct. Jean Lena Annette Conan Doyle (afterwards Air Commandant Dame Jean Conan Doyle, Lady Bromet) born.

1913   (Feb.) Writes 'Great Britain and the Next War' (*Fortnightly Review*). (Aug.) Second Challenger story, *The Poison Belt*, published. (Dec.) 'The Dying Detective' published in the *Strand*. ACD campaigns for a channel tunnel.

1914   (July) 'Danger!', warning of the dangers of a war-time blockade of Britain, published in the *Strand*. (4 Aug.) Britain declares war on Germany; ACD forms local volunteer force.

1914–15   (Sept.) *The Valley of Fear* begins serialization in the *Strand* (concluding May 1915).

1915   (27 Feb.) *The Valley of Fear* published by George H. Doran in New York. (June) *The Valley of Fear* published in London by Smith, Elder (transferred with rest of ACD stock to John Murray when the firm is sold on the death of Reginald Smith). Five Holmes films released in Germany (ten more during the war).

1916   (Apr., May) First instalments of *The British Campaign in France and Flanders 1914* appear in the *Strand*. (Aug.) *A Visit to Three Fronts* published. Sir Roger Casement convicted of high treason after Dublin Easter Week Rising and executed despite appeals for clemency by ACD and others.

1917   War censor interdicts ACD's history of the 1916 campaigns in the *Strand*. (Sept.) 'His Last Bow' published in the *Strand*. (Oct.) *His Last Bow* published by John Murray (includes 'The Cardboard Box').

1918   (Apr.) ACD publishes *The New Revelation*, proclaiming himself a Spiritualist. (Dec.) *Danger! and Other Stories* published. Permitted to resume accounts of 1916 and 1917 campaigns in the *Strand*, but that for 1918 never serialized. Death of eldest son, Captain Kingsley Conan Doyle, from influenza aggravated by war wounds.

1919   Death of Brigadier-General Innes Doyle, from post-war pneumonia.

1920–30   ACD engaged in world-wide crusade for Spiritualism.

1921–2   ACD's one-act play, *The Crown Diamond*, tours with Dennis Neilson-Terry as Holmes.

1921   (Oct.) 'The Mazarin Stone' (apparently based on *The Crown Diamond*) published in the *Strand*. Death of mother, Mary Foley Doyle.

1922   (Feb.–Mar.) 'The Problem of Thor Bridge' in the *Strand*. (July) John Murray publishes a collected edition of the non-Holmes short stories in six volumes: *Tales of the Ring and the Camp*, *Tales of Pirates and Blue Water*, *Tales of Terror and Mystery*, *Tales of Twilight and the Unseen*, *Tales of Adventure and Medical Life*, and (Nov.) *Tales of Long Ago*. (Sept.) Collected edition of ACD's *Poems* published by Murray.

1923   (Mar.) 'The Creeping Man' published in the *Strand*.

1924   (Jan.) 'The Sussex Vampire' appears in the *Strand*. (June) 'How Watson Learned the Trick', ACD's own Holmes pastiche, appears in *The Book of the Queen's Dolls' House Library*. (Sept.) *Memories and Adventures* published (reprinted with additions and deletions 1930).

1925   (Jan.) 'The Three Garridebs' and (Feb.–Mar.) 'The Illustrious Client' published in the *Strand*. (July) *The Land of Mist*, a Spiritualist novel featuring Challenger, begins serialization in the *Strand*.

1926   (Mar.) *The Land of Mist* published. *Strand* publishes 'The Three Gables' (Oct.), 'The Blanched Soldier' (Nov.), and 'The Lion's Mane' (Dec.).

1927   *Strand* publishes 'The Retired Colourman' (Jan.), 'The Veiled Lodger' (Feb.), and 'Shoscombe Old Place' (Apr.). (June) Murray publishes *The Case-Book of Sherlock Holmes*.

1928   (Oct.) *The Complete Sherlock Holmes Short Stories* published by Murray.

1929   (June) *The Conan Doyle Stories* (containing the six separate volumes issued by Murray in 1922) published. (July) *The Maracot Deep and Other Stories*, ACD's last collection of his fictional work.

1930   (7 July, 8.30 a.m.) Death of Arthur Conan Doyle. 'Education never ends, Watson. It is a series of lessons with the greatest for the last' ('The Red Circle').

The Memoirs of
Sherlock Holmes

Silver Blaze

'I AM afraid, Watson, that I shall have to go,' said Holmes, as we sat down together to our breakfast one morning.

'Go! Where to?'

'To Dartmoor—to King's Pyland.'

I was not surprised. Indeed, my only wonder was that he had not already been mixed up in this extraordinary case, which was the one topic of conversation through the length and breadth of England. For a whole day my companion had rambled about the room with his chin upon his chest and his brows knitted, charging and re-charging his pipe with the strongest black tobacco, and absolutely deaf to any of my questions or remarks. Fresh editions of every paper had been sent up by our newsagent only to be glanced over and tossed down into a corner. Yet, silent as he was, I knew perfectly well what it was over which he was brooding. There was but one problem before the public which could challenge his powers of analysis, and that was the singular disappearance of the favourite for the Wessex Cup, and the tragic murder of its trainer. When, therefore, he suddenly announced his intention of setting out for the scene of the drama, it was only what I had both expected and hoped for.

'I should be most happy to go down with you if I should not be in the way,' said I.

'My dear Watson, you would confer a great favour upon me by coming. And I think that your time will not be mis-spent, for there are points about this case which promise to make it an absolutely unique one. We have, I think, just time to catch our train at Paddington,* and I will go further into the matter upon our journey. You would oblige me by bringing with you your very excellent field-glass.'

And so it happened that an hour or so later I found myself in the corner of a first-class carriage, flying along, *en route* for

Exeter, while Sherlock Holmes, with his sharp, eager face framed in his ear-flapped travelling cap,* dipped rapidly into the bundle of fresh papers which he had procured at Paddington. We had left Reading far behind us before he thrust the last of them under the seat, and offered me his cigar-case.

'We are going well,' said he, looking out of the window and glancing at his watch. 'Our rate at present is fifty-three and a half miles an hour.'

'I have not observed the quarter-mile posts,' said I.

'Nor have I. But the telegraph posts upon this line are sixty yards apart, and the calculation is a simple one.* I presume that you have already looked into this matter of the murder of John Straker and the disappearance of Silver Blaze?'

'I have seen what the *Telegraph* and the *Chronicle** have to say.'

'It is one of those cases where the art of the reasoner should be used rather for the sifting of details than for the acquiring of fresh evidence. The tragedy has been so uncommon, so complete, and of such personal importance to so many people, that we are suffering from a plethora of surmise, conjecture, and hypothesis. The difficulty is to detach the framework of fact—of absolute, undeniable fact —from the embellishments of theorists and reporters. Then, having established ourselves upon this sound basis, it is our duty to see what inferences may be drawn, and which are the special points upon which the whole mystery turns. On Tuesday evening I received telegrams, both from Colonel Ross, the owner of the horse, and from Inspector Gregory, who is looking after the case, inviting my co-operation.'

'Tuesday evening!' I exclaimed. 'And this is Thursday morning. Why did you not go down yesterday?'

'Because I made a blunder, my dear Watson—which is, I am afraid, a more common occurrence than anyone would think who only knew me through your memoirs. The fact is, that I could not believe it possible that the most remarkable horse in England could long remain concealed, especially in so sparsely inhabited a place as the north of Dartmoor. From hour to hour yesterday I expected to hear that he had been found, and that his abductor was the murderer of John Straker. When, however, another morning had come and I found that, beyond the arrest of young Fitzroy Simpson,* nothing had been done, I felt that it was time for me to take action. Yet in some ways I feel that yesterday has not been wasted.'

'You have formed a theory, then?'

'At least I have a grip of the essential facts of the case. I shall enumerate them to you, for nothing clears up a case so much as stating it to another person, and I can hardly expect your co-operation if I do not show you the position from which we start.'

I lay back against the cushions, puffing at my cigar, while Holmes, leaning forward, with his long thin fore-finger checking off the points upon the palm of his left hand, gave me a sketch of the events which had led to our journey.

'Silver Blaze,' said he, 'is from the Isonomy stock,* and holds as brilliant a record as his famous ancestor. He is now in his fifth year, and has brought in turn each of the prizes of the turf to Colonel Ross, his fortunate owner. Up to the time of the catastrophe he was first favourite for the Wessex Cup, the betting being three to one on. He has always, however, been a prime favourite with the racing public, and has never yet disappointed them, so that even at short odds enormous sums of money have been laid upon him. It is obvious, therefore, that there were many people who had the strongest interest in preventing Silver Blaze from being there at the fall of the flag, next Tuesday.

'This fact was, of course, appreciated at King's Pyland, where the Colonel's training stable is situated. Every precaution was taken to guard the favourite. The trainer, John Straker, is a retired jockey, who rode in Colonel Ross's colours before he became too heavy for the weighing chair. He has served the Colonel for five years as jockey, and for seven as trainer, and has always shown himself to be a zealous and honest servant. Under him were three lads, for the establishment was a small one, containing only four horses in all. One of these lads sat up each night in the stable,* while the others slept in the loft. All three bore excellent characters. John Straker, who is a married man, lived in a small villa about two hundred yards from the stables. He has no children, keeps one maid-servant, and is comfortably off. The country round is very lonely, but about half a mile to the north there is a small cluster of villas which have been built by a Tavistock contractor for the use of invalids and others who may wish to enjoy the pure Dartmoor air. Tavistock itself lies two miles to the west, while across the moor, also about two miles distant, is the larger training establishment of Capleton,* which belongs to Lord Backwater, and is managed by Silas Brown. In every other direction the moor is a complete wilderness, inhabited only by a few roaming gipsies.* Such was the general situation last Monday night, when the catastrophe occurred.

'On that evening the horses had been exercised and watered as usual, and the stables were locked up at nine o'clock. Two of the lads walked up to the trainer's house, where they had supper in the kitchen, while the third, Ned Hunter, remained on guard. At a few minutes after nine the maid, Edith Baxter, carried down to the stables his supper, which consisted of a dish of curried mutton. She took no liquid, as there was a water-tap in the stables, and it was the rule that the lad on duty should drink nothing else. The maid carried a lantern with her, as it was very dark, and the path ran across the open moor.

'Edith Baxter was within thirty yards of the stables when a man appeared out of the darkness and called to her to stop. As he stepped into the circle of yellow light thrown by the lantern she saw that he was a person of gentlemanly bearing, dressed in a grey suit of tweed with a cloth cap. He wore gaiters, and carried a heavy stick with a knob to it. She was most impressed, however, by the extreme pallor of his face and by the nervousness of his manner. His age, she thought, would be rather over thirty than under it.

' "Can you tell me where I am?" he asked. "I had almost made up my mind to sleep on the moor, when I saw the light of your lantern."

' "You are close to the King's Pyland training stables," she said.

' "Oh, indeed! What a stroke of luck!" he cried. "I understand that a stable boy sleeps there alone every night. Perhaps that is his supper which you are carrying to him. Now I am sure that you would not be too proud to earn the price of a new dress, would you?" He took a piece of white paper folded up out of his waistcoat pocket. "See that the boy has this to-night, and you shall have the prettiest frock that money can buy."

'She was frightened by the earnestness of his manner, and ran past him to the window through which she was accustomed to hand the meals. It was already open, and Hunter was seated at the small table inside. She had begun to tell him of what had happened, when the stranger came up again.

' "Good evening," said he, looking through the window, "I wanted to have a word with you." The girl has sworn that as he spoke she noticed the corner of the little paper packet protruding from his closed hand.

' "What business have you here?" asked the lad.

' "It's business that may put something into your pocket," said the other. "You've two horses in for the Wessex Cup—Silver Blaze and Bayard. Let me have the straight tip, and you won't be a loser. Is it a fact that at the weights Bayard could give the other a hundred yards in five furlongs,* and that the stable have put their money on him?"

' "So you're one of those damned touts,*" cried the lad. "I'll show you how we serve them in King's Pyland." He sprang up and rushed across the stable to unloose the dog. The girl fled away to the house, but as she ran she looked back, and saw that the stranger was leaning through the window. A minute later, however, when Hunter rushed out with the hound he was gone, and though the lad ran all round the buildings he failed to find any trace of him.'

7

'One moment!' I asked. 'Did the stable boy, when he ran out with the dog, leave the door unlocked behind him?'

'Excellent, Watson; excellent!' murmured my companion. 'The importance of the point struck me so forcibly that I sent a special wire to Dartmoor yesterday to clear the matter up. The boy locked the door before he left it. The window, I may add, was not large enough for a man to get through.

'Hunter waited until his fellow grooms had returned, when he sent a message up to the trainer and told him what had occurred. Straker was excited at hearing the account, although he does not seem to have quite realized its true significance. It left him, however, vaguely uneasy, and Mrs Straker, waking at one in the morning, found that he was dressing. In reply to her inquiries, he said that he could not sleep on account of his anxiety about the horses, and that he intended to walk down to the stables to see that all was well. She begged him to remain at home, as she could hear the rain pattering against the windows, but in spite of her entreaties he pulled on his large mackintosh and left the house.

'Mrs Straker awoke at seven in the morning, to find that her husband had not yet returned. She dressed herself hastily, called the maid, and set off for the stables. The door was open; inside, huddled together upon a chair, Hunter was sunk in a state of absolute stupor, the favourite's stall was empty, and there were no signs of his trainer.

'The two lads who slept in the chaff-cutting loft above the harness-room were quickly aroused. They had heard nothing during the night, for they are both sound sleepers. Hunter was obviously under the influence of some powerful drug; and, as no sense could be got out of him, he was left to sleep it off while the two lads and the two women ran out in search of the absentees. They still had hopes that the trainer had for some reason taken out the horse for early exercise, but on ascending the knoll near the house, from which all the neighbouring moors were visible, they not only could see no signs of the favourite, but they perceived something which warned them that they were in the presence of a tragedy.

'About a quarter of a mile from the stables, John Straker's overcoat was flapping from a furze bush. Immediately beyond there was a bowl-shaped depression in the moor, and at the bottom of this was found the dead body of the unfortunate trainer. His head had been shattered by a savage blow from some heavy weapon, and he was wounded in the thigh, where there was a long, clean cut, inflicted evidently by some very sharp instrument. It was clear, however, that Straker had defended himself vigorously against his assailants, for in his right hand he held a small knife, which was clotted with blood up to the handle, while in his left he grasped a red and black silk cravat, which was recognized by the maid as having been worn on the preceding evening by the stranger who had visited the stables.

'Hunter, on recovering from his stupor, was also quite positive as to the ownership of the cravat. He was equally certain that the same stranger had, while standing at the window, drugged his curried mutton, and so deprived the stables of their watchman.

'As to the missing horse, there were abundant proofs in the mud which lay at the bottom of the fatal hollow that he had been there at the time of the struggle. But from that morning he has disappeared; and, although a large reward has been offered, and all the gipsies of Dartmoor are on the alert, no news has come of him. Finally, an analysis has shown that the remains of his supper, left by the stable lad, contain an appreciable quantity of powdered opium, while the people of the house partook of the same dish on the same night without any ill effect.

'Those are the main facts of the case stripped of all surmise and stated as baldly as possible. I shall now recapitulate what the police have done in the matter.

'Inspector Gregory, to whom the case has been committed, is an extremely competent officer. Were he but gifted with imagination he might rise to great heights in his profession. On his arrival he promptly found and arrested the man upon whom suspicion naturally rested. There was little difficulty in finding him, for he was thoroughly well known in the neighbourhood. His name, it appears, was Fitzroy Simpson. He was a man of excellent birth and education, who had squandered a fortune upon the turf, and who lived now by doing a little quiet and genteel book-making in the sporting clubs of London. An examination of his betting-book shows that bets to the amount of five thousand pounds had been registered by him against the favourite.

'On being arrested, he volunteered the statement that he had come down to Dartmoor in the hope of getting some information about the King's Pyland horses, and also about Desborough, the second favourite, which was in charge of Silas Brown, at the Capleton stables. He did not attempt to deny that he had acted as described upon the evening before, but declared that he had no sinister designs, and had simply wished to obtain first-hand information. When confronted with the cravat he turned very pale, and was utterly unable to account for its presence in the hand of the murdered man. His wet clothing showed that he had been out in the storm of the night before, and his stick, which was a Penang lawyer,* weighted with lead, was just such a weapon as might, by repeated blows, have inflicted the terrible injuries to which the trainer had succumbed.

'On the other hand, there was no wound upon his person, while the state of Straker's knife would show that one, at least, of his assailants must bear his mark upon him. There you have it all in a nutshell, Watson, and if you can give me any light I shall be infinitely obliged to you.'

I had listened with the greatest interest to the statement which Holmes, with characteristic clearness, had laid before me. Though most of the facts were familiar to me, I had not sufficiently appreciated their relative importance, nor their connection with each other.

'Is it not possible,' I suggested, 'that the incised wound upon Straker may have been caused by his own knife in the convulsive struggles which follow any brain injury?'

'It is more than possible; it is probable,' said Holmes. 'In that case, one of the main points in favour of the accused disappears.'

'And yet,' said I, 'even now I fail to understand what the theory of the police can be.'

'I am afraid that whatever theory we state has very grave objections to it,' returned my companion. 'The police imagine, I take it, that this Fitzroy Simpson, having drugged the lad, and having in some way obtained a duplicate key, opened the stable door, and took out the horse, with the intention, apparently, of kidnapping him altogether. His bridle is missing, so that Simpson must have put it on. Then, having left the door open behind him, he was leading the horse away over the moor, when he was either met or overtaken by the trainer. A row naturally ensued, Simpson beat out the trainer's brains with his heavy stick without receiving any injury from the small knife which Straker used in self-defence, and then the thief either led the horse on to some secret hiding-place, or else it may have bolted during the struggle, and be now wandering out on the moors. That is the case as it appears to the police, and improbable as it is, all other explanations are more improbable still. However, I shall very quickly test the matter when I am once upon the spot, and until then I really cannot see how we can get much further than our present position.'

It was evening before we reached the little town of Tavistock, which lies, like the boss of a shield, in the middle of the huge circle of Dartmoor. Two gentlemen were awaiting us at the station; the one a tall, fair man with lion-like hair and beard, and curiously penetrating light-blue eyes, the other a small alert person, very neat and dapper, in a frock-coat and gaiters, with trim little side-whiskers and an eye-glass. The latter was Colonel Ross, the well-known sportsman, the other Inspector Gregory, a man who was rapidly making his name in the English detective service.

'I am delighted that you have come down, Mr Holmes,' said the Colonel. 'The Inspector here has done all that could possibly be suggested; but I wish to leave no stone unturned in trying to avenge poor Straker, and in recovering my horse.'

'Have there been any fresh developments?' asked Holmes.

'I am sorry to say that we have made very little progress,' said the Inspector. 'We have an open carriage outside, and as you would no doubt like to see the place before the light fails, we might talk it over as we drive.'

A minute later we were all seated in a comfortable landau* and were rattling through the quaint old Devonshire town. Inspector Gregory was full of his case, and poured out a stream of remarks, while Holmes threw in an occasional question or interjection. Colonel Ross leaned back with his arms folded and his hat tilted over his eyes, while I listened with interest to the dialogue of the two detectives. Gregory was formulating his theory, which was almost exactly what Holmes had foretold in the train.

'The net is drawn pretty close round Fitzroy Simpson.' he remarked, 'and I believe myself that he is our man. At the same time, I recognize that the evidence is purely circumstantial, and that some new development may upset it.'

'How about Straker's knife?'

'We have quite come to the conclusion that he wounded himself in his fall.'

'My friend Dr Watson made that suggestion to me as we came down. If so, it would tell against this man Simpson.'

'Undoubtedly. He has neither a knife nor any sign of a wound. The evidence against him is certainly very strong. He had a great interest in the disappearance of the favourite, he lies under the suspicion of having poisoned the stable-boy, he was undoubtedly out in the storm, he was armed with a heavy stick, and his cravat was found in the dead man's hand. I really think we have enough to go before a jury.'

Holmes shook his head. 'A clever counsel would tear it all to rags,' said he. 'Why should he take the horse out of the stable? If he wished to injure it, why could he not do it there? Has a duplicate key been found in his possession? What chemist sold him the powdered opium? Above all, where could he, a stranger to the district, hide a horse, and such a horse as this? What is his own explanation as to the paper which he wished the maid to give to the stable-boy?'

'He says that it was a ten-pound note. One was found in his purse. But your other difficulties are not so formidable as they seem. He is not a stranger to the district. He has twice lodged at Tavistock in the summer. The opium was probably brought from London. The key, having served its purpose, would be hurled away. The horse may lie at the bottom of one of the pits or old mines upon the moor.'

'What does he say about the cravat?'

'He acknowledges that it is his, and declares that he had lost it. But a new element has been introduced into the case which may account for his leading the horse from the stable.'

Holmes pricked up his ears.

'We have found traces which show that a party of gipsies encamped on Monday night within a mile of the spot where the murder took place. On Tuesday they were gone. Now, presuming that there was some understanding between Simpson and these gipsies, might he not have been leading the horse to them when he was overtaken, and may they not have him now?'

'It is certainly possible.'

'The moor is being scoured for these gipsies. I have also examined every stable and outhouse in Tavistock, and for a radius of ten miles.'

'There is another training stable quite close, I understand?'

'Yes, and that is a factor which we must certainly not neglect. As Desborough, their horse, was second in the betting, they had an interest in the disappearance of the favourite. Silas Brown, the trainer, is known to have had large bets upon the event, and he was no friend to poor Straker. We have, however, examined the stables, and there is nothing to connect him with the affair.'

'And nothing to connect this man Simpson with the interests of the Capleton stables?'

'Nothing at all.'

Holmes leaned back in the carriage and the conversation ceased. A few minutes later our driver pulled up at a neat little red-brick villa with overhanging eaves, which stood by the road. Some distance off, across a paddock, lay a long grey-tiled out-building. In every other direction the low curves of the moor, bronze-coloured from the fading ferns, stretched away to the sky-line, broken only by the steeples of Tavistock, and by a cluster of houses away to the westward, which marked the Capleton stables. We all sprang out, with the exception of Holmes, who continued to lean back with his eyes fixed upon the sky in front of him, entirely absorbed in his own thoughts. It was only when I touched his arm that he roused himself with a violent start and stepped out of the carriage.

'Excuse me,' said he, turning to Colonel Ross, who had looked at him in some surprise. 'I was day-dreaming.' There was a gleam in his eyes and a suppressed excitement in his manner which convinced me, used as I was to his ways, that his hand was upon a clue, though I could not imagine where he had found it.

'Perhaps you would prefer at once to go on to the scene of the crime, Mr Holmes?' said Gregory.

'I think that I should prefer to stay here a little and go into one or two questions of detail. Straker was brought back here, I presume?'

'Yes, he lies upstairs. The inquest is to-morrow.'

'He has been in your service some years, Colonel Ross?'

'I have always found him an excellent servant.'

'I presume that you made an inventory of what he had in his pockets at the time of his death, Inspector?'

'I have the things themselves in the sitting-room if you would care to see them.'

'I should be very glad.'

We all filed into the front room and sat round the central table, while the Inspector unlocked a square tin box and laid a small heap of things before us. There was a box of vestas, two inches of tallow candle, an A.D.P. briar-root pipe,* a pouch of sealskin with half an ounce of long-cut Cavendish,* a silver watch with a gold chain, five sovereigns in gold, an aluminium pencil-case, a few papers, and an ivory-handled knife with a very delicate inflexible blade marked 'Weiss and Co., London.'

'This is a very singular knife,' said Holmes, lifting it up and examining it minutely. 'I presume, as I see bloodstains upon it, that it is the one which was found in the dead man's grasp. Watson, this knife is surely in your line.'

'It is what we call a cataract knife,*' said I.

'I thought so. A very delicate blade devised for very delicate work. A strange thing for a man to carry with him upon a rough expedition, especially as it would not shut in his pocket.'

'The tip was guarded by a disc of cork which we found beside his body,' said the Inspector. 'His wife tells us that the knife had lain for some days upon the dressing-table, and that he had picked it up as he left the room. It was a poor weapon, but perhaps the best that he could lay his hand on at the moment.'

'Very possibly. How about these papers?'

'Three of them are receipted hay-dealers' accounts. One of them is a letter of instructions from Colonel Ross. This other is a milliner's account for thirty-seven pounds fifteen, made out by Madame Lesurier, of Bond Street, to William Darbyshire. Mrs Straker tells us that Darbyshire was a friend of her husband's, and that occasionally his letters were addressed here.'

'Madame Darbyshire had somewhat expensive tastes,' remarked Holmes, glancing down the account. 'Twenty-two guineas* is rather heavy for a single costume. However, there appears to be nothing more to learn, and we may now go down to the scene of the crime.'

As we emerged from the sitting-room a woman who had been waiting in the passage took a step forward and laid her hand upon the Inspector's sleeve. Her face was haggard, and thin, and eager; stamped with the print of a recent horror.

'Have you got them? Have you found them?' she panted.

'No, Mrs Straker; but Mr Holmes, here, has come from London to help us, and we shall do all that is possible.'

'Surely I met you in Plymouth, at a garden party, some little time ago, Mrs Straker,' said Holmes.

'No, sir; you are mistaken.'

'Dear me; why, I could have sworn to it. You wore a costume of dove-coloured silk with ostrich feather trimming.'

'I never had such a dress, sir,' answered the lady.

'Ah; that quite settles it,' said Holmes; and, with an apology, he followed the Inspector outside. A short walk across the moor took us to the hollow in which the body had been found. At the brink of it was the furze bush upon which the coat had been hung.

'There was no wind that night, I understand,' said Holmes.

'None; but very heavy rain.'

'In that case the overcoat was not blown against the furze bushes, but placed there.'

'Yes, it was laid across the bush.'

'You fill me with interest. I perceive that the ground has been trampled up a good deal. No doubt many feet have been there since Monday night.'

'A piece of matting has been laid here at the side, and we have all stood upon that.'

'Excellent.'

'In this bag I have one of the boots which Straker wore, one of Fitzroy Simpson's shoes, and a cast horseshoe of Silver Blaze.'

'My dear Inspector, you surpass yourself!'

Holmes took the bag, and descending into the hollow he pushed the matting into a more central position. Then stretching himself upon his face and leaning his chin upon his hands he made a careful study of the trampled mud in front of him.

'Hullo!' said he, suddenly, 'what's this?'

It was a wax vesta, half burned, which was so coated with mud that it looked at first like a little chip of wood.

'I cannot think how I came to overlook it,' said the Inspector, with an expression of annoyance.

'It was invisible, buried in the mud. I only saw it because I was looking for it.'

'What! You expected to find it?'

'I thought it not unlikely.' He took the boots from the bag and compared the impressions of each of them with marks upon the ground. Then he clambered up to the rim of the hollow and crawled about among the ferns and bushes.

'I am afraid that there are no more tracks,' said the Inspector, 'I have examined the ground very carefully for a hundred yards in each direction.'

'Indeed!' said Holmes, rising, 'I should not have the impertinence to do it again after what you say. But I should like to take a little walk over the moors before it grows dark, that I may know my ground to-morrow, and I think that I shall put this horseshoe into my pocket for luck.'

Colonel Ross, who had shown some signs of impatience at my companion's quiet and systematic method of work, glanced at his watch.

'I wish you would come back with me, Inspector,' said he. 'There are several points on which I should like your advice, and especially as to whether we do not owe it to the public to remove our horse's name from the entries for the Cup.'

'Certainly not,' cried Holmes, with decision; 'I should let the name stand.'

The Colonel bowed. 'I am very glad to have had your opinion, sir,' said he. 'You will find us at poor Straker's house when you have finished your walk, and we can drive together into Tavistock.'

He turned back with the Inspector, while Holmes and I walked slowly across the moor. The sun was beginning to sink behind the stables of Capleton, and the long sloping plain in front of us was tinged with gold, deepening into rich, ruddy brown where the faded ferns and brambles caught the evening light. But the glories of the landscape were all wasted upon my companion, who was sunk in the deepest thought.

'It's this way, Watson,' he said at last. 'We may leave the question of who killed John Straker for the instant, and confine ourselves to finding out what has become of the horse. Now supposing that he broke away during or after the tragedy, where could he have gone to? The horse is a very gregarious creature. If left to himself his instincts would have been either to return to King's Pyland, or go over to Capleton. Why should he run wild upon the moor? He would surely have been seen by now. And why should gipsies kidnap him? These people always clear out when they hear of trouble, for they do not wish to be pestered by the police. They could not hope to sell such a horse. They would run a great risk and gain nothing by taking him. Surely that is clear.'

'Where is he, then?'

'I have already said that he must have gone to King's Pyland or to Capleton. He is not at King's Pyland, therefore he is at Capleton. Let us take that as a working hypothesis and see what it leads us to. This part of the moor, as the Inspector remarked, is very hard and dry. But it falls away towards Capleton, and you can see from here that there is a long hollow over yonder, which must have been very wet on Monday night. If our supposition is correct, then the horse must have crossed that, and there is the point where we should look for his tracks.'

We had been walking briskly during this conversation, and a few more minutes brought us to the hollow in question. At Holmes's request I walked down the bank to the right, and he to the left, but I had not taken fifty paces before I heard him give a shout, and saw him waving his hand to me. The track of a horse was plainly outlined in the soft earth in front of him, and the shoe which he took from his pocket exactly fitted the impression.

'See the value of imagination,' said Holmes. 'It is the one quality which Gregory lacks. We imagined what might have happened, acted upon the supposition, and find ourselves justified. Let us proceed.'

We crossed the marshy bottom and passed over a quarter of a mile of dry, hard turf. Again the ground sloped and again we came on the tracks. Then we lost them for half a mile, but only to pick them up once more quite close to Capleton. It was Holmes who saw them first, and he stood pointing with a look of triumph upon his face. A man's track was visible beside the horse's.

'The horse was alone before,' I cried.

'Quite so. It was alone before. Hullo! what is this?'

The double track turned sharp off and took the direction of King's Pyland. Holmes whistled, and we both followed along after it. His eyes were on the trail, but I happened to look a little to one side, and saw to my surprise the same tracks coming back again in the opposite direction.

'One for you, Watson,' said Holmes, when I pointed it out; 'you have saved us a long walk which would have brought us back on our own traces. Let us follow the return track.'

We had not to go far. It ended at the paving of asphalt which led up to the gates of the Capleton stables. As we approached a groom ran out from them.

'We don't want any loiterers about here,' said he.

'I only wish to ask a question,' said Holmes, with his finger and thumb in his waistcoat pocket. 'Should I be too early to see your master, Mr Silas Brown, if I were to call at five o'clock to-morrow morning?'

'Bless you, sir, if anyone is about he will be, for he is always the first stirring. But here he is, sir, to answer your questions for himself. No, sir, no; it's as much as my place is worth to let him see me touch your money. Afterwards, if you like.'

As Sherlock Holmes replaced the half-crown which he had drawn from his pocket, a fierce-looking, elderly man strode out from the gate with a hunting-crop swinging in his hand.

'What's this, Dawson?' he cried. 'No gossiping! Go about your business! And you—what the devil do you want here?'

'Ten minutes' talk with you, my good sir,' said Holmes, in the sweetest of voices.

'I've no time to talk to every gadabout. We want no strangers here. Be off, or you may find a dog at your heels.'

Holmes leaned forward and whispered something in the trainer's ear. He started violently and flushed to the temples.

'It's a lie!' he shouted. 'An infernal lie!'

'Very good! Shall we argue about it here in public, or talk it over in your parlour?'

'Oh, come in if you wish to.'

Holmes smiled. 'I shall not keep you more than a few minutes, Watson,' he said. 'Now, Mr Brown, I am quite at your disposal.'

It was quite twenty minutes, and the reds had all faded into greys before Holmes and the trainer reappeared. Never have I seen such a change as had been brought about in Silas Brown in that short time. His face was ashy pale, beads of perspiration shone upon his brow, and his hands shook until the hunting-crop wagged like a branch in the wind. His bullying, overbearing manner was all gone too, and he cringed along at my companion's side like a dog with its master.

'Your instructions will be done. It shall be done,' said he.

'There must be no mistake,' said Holmes, looking round at him. The other winced as he read the menace in his eyes.

'Oh, no, there shall be no mistake. It shall be there. Should I change it first or not?'

Holmes thought a little and then burst out laughing. 'No, don't,' said he. 'I shall write to you about it. No tricks now or—'

'Oh, you can trust me, you can trust me!'

'You must see to it on the day as if it were your own.'

'You can rely upon me.'*

'Yes, I think I can. Well, you shall hear from me to-morrow.' He turned upon his heel, disregarding the trembling hand which the other held out to him, and we set off for King's Pyland.

'A more perfect compound of the bully, coward, and sneak than Master Silas Brown I have seldom met with,' remarked Holmes, as we trudged along together.

'He has the horse, then?'

'He tried to bluster out of it, but I described to him so exactly what his actions had been upon that morning, that he is convinced that I was watching him. Of course, you observed the peculiarly square toes in the impressions, and that his own boots exactly corresponded to them. Again, of course, no subordinate would have dared to have done such a thing. I described to him how when, according to his custom, he was the first down, he perceived a strange horse wandering over the moor; how he went out to it, and his astonishment at recognising from the white forehead which has given the favourite its name that chance had put in his power the only horse which could beat the one upon which he had put his money. Then I described how his first impulse had been to lead it back to King's Pyland, and how the devil had shown him how he could hide the horse until the race was over, and how he had led it back and concealed it at Capleton. When I told him every detail he gave it up, and thought only of saving his own skin.'

'But his stables had been searched.'

'Oh, an old horse-faker like him has many a dodge.'

'But are you not afraid to leave the horse in his power now, since he has every interest in injuring it?'

'My dear fellow, he will guard it as the apple of his eye. He knows that his only hope of mercy is to produce it safe.'

'Colonel Ross did not impress me as a man who would be likely to show much mercy in any case.'

'The matter does not rest with Colonel Ross. I follow my own methods, and tell as much or as little as I choose. That is the advantage of being unofficial. I don't know whether you observed it, Watson, but the Colonel's manner has been just a trifle cavalier to me. I am inclined now to have a little amusement at his expense. Say nothing to him about the horse.'

'Certainly not, without your permission.'

'And, of course, this is all quite a minor case compared with the question of who killed John Straker.'

'And you will devote yourself to that?'

'On the contrary, we both go back to London by the night train.'

I was thunderstruck by my friend's words. We had only been a few hours in Devonshire, and that he should give up an investigation which he had begun so brilliantly was quite incomprehensible to me. Not a word more could I draw from him until we were back at the trainer's house. The Colonel and the Inspector were awaiting us in the parlour.

'My friend and I return to town by the midnight express,' said Holmes. 'We have had a charming little breath of your beautiful Dartmoor air.'

The Inspector opened his eyes, and the Colonel's lips curled in a sneer.

'So you despair of arresting the murderer of poor Straker,' said he.

Holmes shrugged his shoulders. 'There are certainly grave difficulties in the way,' said he. 'I have every hope, however, that your horse will start upon Tuesday, and I beg that you will have your jockey in readiness. Might I ask for a photograph of Mr John Straker?'

The Inspector took one from an envelope in his pocket and handed it to him.

'My dear Gregory, you anticipate all my wants. If I might ask you to wait here for an instant, I have a question which I should like to put to the maid.'

'I must say that I am rather disappointed in our London consultant,' said Colonel Ross, bluntly, as my friend left the room. 'I do not see that we are any further than when he came.'

'At least, you have his assurance that your horse will run,' said I.

'Yes, I have his assurance,' said the Colonel, with a shrug of his shoulders. 'I should prefer to have the horse.'

I was about to make some reply in defence of my friend, when he entered the room again.

'Now, gentlemen,' said he, 'I am quite ready for Tavistock.'

As we stepped into the carriage one of the stable lads held the door open for us. A sudden idea seemed to occur to Holmes, for he leaned forward and touched the lad upon the sleeve.

'You have a few sheep in the paddock,' he said. 'Who attends to them?'

'I do, sir.'

'Have you noticed anything amiss with them of late?'

'Well, sir, not of much account; but three of them have gone lame, sir.'

I could see that Holmes was extremely pleased, for he chuckled and rubbed his hands together.

'A long shot, Watson; a very long shot!' said he, pinching my arm. 'Gregory, let me recommend to your attention this singular epidemic among the sheep. Drive on, coachman!'

Colonel Ross still wore an expression which showed the poor opinion which he had formed of my companion's ability, but I saw by the Inspector's face that his attention had been keenly aroused.

'You consider that to be important?' he asked.

'Exceedingly so.'

'Is there any other point to which you would wish to draw my attention?'

'To the curious incident of the dog in the night-time.'*

'The dog did nothing in the night-time.'

'That was the curious incident,' remarked Sherlock Holmes.

Four days later Holmes and I were again in the train bound for Winchester, to see the race for the Wessex Cup. Colonel Ross met us, by appointment, outside the station, and we drove in his drag* to the course beyond the town. His face was grave and his manner was cold in the extreme.

'I have seen nothing of my horse,' said he.

'I suppose that you would know him when you saw him?' asked Holmes.

The Colonel was very angry. 'I have been on the turf for twenty years, and never was asked such a question as that before,' said he. 'A child would know Silver Blaze with his white forehead and his mottled off fore-leg.'*

'How is the betting?'

'Well, that is the curious part of it. You could have got fifteen to one yesterday, but the price has become shorter and shorter, until you can hardly get three to one now.'

'Hum!' said Holmes. 'Somebody knows something, that is clear!'

As the drag drew up in the inclosure near the grandstand, I glanced at the card to see the entries. It ran:

Wessex Plate.* 50 sovs. each h ft, with 1,000 sovs. added, for four-and five-year olds. Second £300. Third £200. New course (one mile and five furlongs).

1. Mr Heath Newton's The Negro (red cap, cinnamon jacket).
2. Colonel Wardlaw's Pugilist (pink cap, blue and black jacket).
3. Lord Backwater's Desborough (yellow cap and sleeves).
4. Colonel Ross's Silver Blaze (black cap, red jacket).
5. Duke of Balmoral's* Iris (yellow and black stripes).
6. Lord Singleford's Rasper (purple cap, black sleeves).

'We scratched our other one and put all hopes on your word,' said the Colonel. 'Why, what is that? Silver Blaze favourite?'

'Five to four against Silver Blaze!' roared the ring. 'Five to four against Silver Blaze! Fifteen to five against Desborough! Five to four on the field!'*

'There are the numbers up,' I cried. 'They are all six there.'

'All six there! Then my horse is running,' cried the Colonel in great agitation. 'But I don't see him. My colours have not passed.'

'Only five have passed. This must be he.'

As I spoke a powerful bay horse swept out from the weighing inclosure and cantered past us, bearing on its back the well-known black and red of the Colonel.

'That's not my horse,' cried the owner. 'That beast has not a white hair upon its body. What is this that you have done, Mr Holmes?'

'Well, well, let us see how he gets on,' said my friend, imperturbably. For a few minutes he gazed through my field-glass. 'Capital! An excellent start!' he cried suddenly. 'There they are, coming round the curve!'

From our drag we had a superb view as they came up the straight. The six horses were so close together that a carpet could have covered them, but half way up the yellow of the Capleton stable showed to the front. Before they reached us, however, Desborough's bolt was shot, and the Colonel's horse, coming away with a rush, passed the post a good six lengths before its rival, the Duke of Balmoral's Iris making a bad third.

'It's my race anyhow,' gasped the Colonel, passing his hand over his eyes. 'I confess that I can make neither head nor tail of it. Don't you think that you have kept up your mystery long enough, Mr Holmes?'

'Certainly, Colonel. You shall know everything. Let us all go round and have a look at the horse together. Here he is,' he continued, as we made our way into the weighing inclosure where only owners and their friends find admittance. 'You have only to wash his face and his leg in spirits of wine* and you will find that he is the same old Silver Blaze as ever.'

'You take my breath away!'

'I found him in the hands of a faker, and took the liberty of running him just as he was sent over.'

'My dear sir, you have done wonders. The horse looks very fit and well. It never went better in its life. I owe you a thousand apologies for having doubted your ability. You have done me a great service by recovering my horse. You would do me a greater still if you could lay your hands on the murderer of John Straker.'

'I have done so,' said Holmes, quietly.

The Colonel and I stared at him in amazement. 'You have got him! Where is he, then?'

'He is here.'

'Here! Where?'

'In my company at the present moment.'

The Colonel flushed angrily. 'I quite recognize that I am under obligations to you, Mr Holmes,' said he, 'but I must regard what you have just said as either a very bad joke or an insult.'

Sherlock Holmes laughed. 'I assure you that I have not associated you with the crime, Colonel,' said he; 'the real murderer is standing immediately behind you!'

He stepped past and laid his hand upon the glossy neck of the thoroughbred.

'The horse!' cried both the Colonel and myself.

'Yes, the horse. And it may lessen his guilt if I say that it was done in self-defence, and that John Straker was a man who was entirely unworthy of your confidence. But there goes the bell; and as I stand to win a little on this next race, I shall defer a more lengthy explanation until a more fitting time.'

We had the corner of a Pullman car* to ourselves that evening as we whirled back to London, and I fancy that the journey was a short one to Colonel Ross as well as to myself, as we listened to our companion's narrative of the events which had occurred at the Dartmoor training stables upon that Monday night, and the means by which he had unravelled them.

'I confess,' said he, 'that any theories which I had formed from the newspaper reports were entirely erroneous. And yet there were indications there, had they not been overlaid by other details which concealed their true import. I went to Devonshire with the conviction that Fitzroy Simpson was the true culprit, although, of course, I saw that the evidence against him was by no means complete.

'It was while I was in the carriage, just as we reached the trainer's house, that the immense significance of the curried mutton occurred to me. You may remember that I was distrait, and remained sitting after you had all alighted. I was marvelling in my own mind how I could possibly have overlooked so obvious a clue.'

'I confess,' said the Colonel, 'that even now I cannot see how it helps us.'

'It was the first link in my chain of reasoning. Powdered opium is by no means tasteless. The flavour is not disagreeable, but it is perceptible. Were it mixed with any ordinary dish, the eater would undoubtedly detect it, and would probably eat no more. A curry was exactly the medium which would disguise this taste. By no possible supposition could this stranger, Fitzroy Simpson, have caused curry to be served in the trainer's family that night, and it is surely too monstrous a coincidence to suppose that he happened to come along with powdered opium upon the very night when a dish happened to be served which would disguise the flavour. That is unthinkable. Therefore Simpson becomes eliminated from the case, and our attention centres upon Straker and his wife, the only two people who could have chosen curried mutton for supper that night. The opium was added after the dish was set aside for the stable boy, for the others had the same for supper with no ill effects. Which of them, then, had access to that dish without the maid seeing them?

'Before deciding that question I had grasped the significance of the silence of the dog, for one true inference invariably suggests others. The Simpson incident had shown me that a dog was kept in the stables, and yet, though someone had been in and had fetched out a horse, he had not barked enough to arouse the two lads in the loft. Obviously the midnight visitor was someone whom the dog knew well.

'I was already convinced, or almost convinced, that John Straker went down to the stables in the dead of the night and took out Silver Blaze. For what purpose? For a dishonest one, obviously, or why should he drug his own stable-boy? And yet I was at a loss to know why. There have been cases before now where trainers have made sure of great sums of money by laying against their own horses, through agents, and then prevented them from winning by fraud. Sometimes it is a pulling jockey. Sometimes it is some surer and subtler means. What was it here? I hoped that the contents of his pockets might help me to form a conclusion.

'And they did so. You cannot have forgotten the singular knife which was found in the dead man's hand, a knife which certainly no sane man would choose for a weapon. It was, as Dr Watson told us, a form of knife which is used for the most delicate operations known in surgery.* And it was to be used for a delicate operation that night. You must know, with your wide experience of turf matters, Colonel Ross, that it is possible to make a slight nick upon the tendons of a horse's ham, and to do it subcutaneously* so as to leave absolutely no trace. A horse so treated would develop a slight lameness which would be put down to a strain in exercise or a touch of rheumatism, but never to foul play.'

'Villain! Scoundrel!' cried the Colonel.

'We have here the explanation of why John Straker wished to take the horse out on to the moor. So spirited a creature would have certainly roused the soundest of sleepers when it felt the prick of the knife. It was absolutely necessary to do it in the open air.'

'I have been blind!' cried the Colonel. 'Of course, that was why he needed the candle, and struck the match.'

'Undoubtedly. But in examining his belongings, I was fortunate enough to discover, not only the method of the crime, but even its motives. As a man of the world, Colonel, you know that men do not carry other people's bills about in their pockets. We have most of us quite enough to do to settle our own. I at once concluded that Straker was leading a double life, and keeping a second establishment. The nature of the bill showed that there was a lady in the case, and one who had expensive tastes. Liberal as you are with your servants, one hardly expects that they can buy twenty-guinea walking dresses for their women. I questioned Mrs Straker as to the dress without her knowing it, and having satisfied myself that it had never reached her, I made a note of the milliner's address, and felt that by calling there with Straker's photograph, I could easily dispose of the mythical Darbyshire.

'From that time on all was plain. Straker had led out the horse to a hollow where his light would be invisible. Simpson, in his flight, had dropped his cravat, and Straker had picked it up with some idea, perhaps, that he might use it in securing the horse's leg. Once in the hollow he had got behind the horse, and had struck a light, but the creature, frightened at the sudden glare, and with the strange instinct of animals feeling that some mischief was intended, had lashed out, and the steel shoe had struck Straker full on the forehead. He had already, in spite of the rain, taken off his overcoat in order to do his delicate task, and so, as he fell, his knife gashed his thigh. Do I make it clear?'

'Wonderful!' cried the Colonel. 'Wonderful! You might have been there.'

'My final shot was, I confess, a very long one. It struck me that so astute a man as Straker would not undertake this delicate tendon-nicking without a little practice. What could he practise on? My eyes fell upon the sheep, and I asked a question which, rather to my surprise, showed that my surmise was correct.'

'You have made it perfectly clear, Mr Holmes.'

'When I returned to London I called upon the milliner, who at once recognized Straker as an excellent customer, of the name of Darbyshire, who had a very dashing wife with a strong partiality for expensive dresses. I have no doubt that this woman had plunged him over head and ears in debt, and so led him into this miserable plot.'

'You have explained all but one thing,' cried the Colonel. 'Where was the horse?'

'Ah, it bolted, and was cared for by one of your neighbours. We must have an amnesty in that direction, I think. This is Clapham Junction, if I am not mistaken, and we shall be in Victoria in less than ten minutes. If you care to smoke a cigar in our rooms, Colonel, I shall be happy to give you any other details which might interest you.'

The Cardboard Box

IN choosing a few typical cases which illustrate the remarkable mental qualities of my friend, Sherlock Holmes, I have endeavoured, as far as possible, to select those which presented the minimum of sensationalism, while offering a fair field for his talents. It is, however, unfortunately, impossible to entirely separate the sensational from the criminal, and a chronicler is left in the dilemma that he must either sacrifice details which are essential to his statement, and so give a false impression of the problem, or he must use matter which chance, and not choice, has provided him with. With this short preface I shall turn to my notes of what proved to be a strange, though a peculiarly terrible, chain of events.

It was a blazing hot day in August. Baker Street was like an oven, and the glare of the sunlight upon the yellow brickwork of the houses across the road was painful to the eye. It was hard to believe that these were the same walls which loomed so gloomily through the fogs of winter. Our blinds were half-drawn, and Holmes lay curled upon the sofa, reading and re-reading a letter which he had received by the morning post. For myself, my term of service in India had trained me to stand heat better than cold, and a thermometer at 90 was no hardship. But the morning paper was uninteresting. Parliament had risen.* Everybody was out of town, and I yearned for the glades of the New Forest* or the shingle of Southsea.* A depleted bank account had caused me to postpone my holiday, and as to my companion, neither the country nor the sea presented the slightest attraction to him. He loved to lie in the very centre of five millions of people, with his filaments stretching out and running through them, responsive to every little rumour or suspicion of unsolved crime. Appreciation of Nature found no place among his many gifts, and his only change was when he turned his mind from the evil-doer of the town to track down his brother of the country.

Finding that Holmes was too absorbed for conversation I had tossed aside the barren paper and, leaning back in my chair, I fell into a brown study. Suddenly my companion's voice broke in upon my thoughts.

'You are right, Watson,' said he. 'It does seem a most preposterous way of settling a dispute.'

'Most preposterous!' I exclaimed, and then suddenly realizing how he had echoed the inmost thought of my soul, I sat up in my chair and stared at him in blank amazement.

'What is this, Holmes?' I cried. 'This is beyond anything which I could have imagined.'

He laughed heartily at my perplexity.

'You remember,' said he, 'that some little time ago when I read you the passage in one of Poe's sketches* in which a close reasoner follows the unspoken thoughts of his companion, you were inclined to treat the matter as a mere *tour-de-force** of the author. On my remarking that I was constantly in the habit of doing the same thing you expressed incredulity.'

'Oh, no!'

'Perhaps not with your tongue, my dear Watson, but certainly with your eyebrows. So when I saw you throw down your paper and enter upon a train of thought, I was very happy to have the opportunity of reading it off, and eventually of breaking into it, as a proof that I had been in rapport* with you.'

But I was still far from satisfied. 'In the example which you read to me,' said I, 'the reasoner drew his conclusions from the actions of the man whom he observed. If I remember right, he stumbled over a heap of stones, looked up at the stars, and so on. But I have been seated quietly in my chair, and what clues can I have given you?'

'You do yourself an injustice. The features are given to man as the means by which he shall express his emotions, and yours are faithful servants.'

'Do you mean to say that you read my train of thoughts from my features?'

'Your features, and especially your eyes. Perhaps you cannot yourself recall how your reverie commenced?'

'No, I cannot.'

'Then I will tell you. After throwing down your paper, which was the action which drew my attention to you, you sat for half a minute with a vacant expression. Then your eyes fixed themselves upon your newly-framed picture of General Gordon,* and I saw by the alteration in your face that a train of thought had been started. But it did not lead very far. Your eyes flashed across to the unframed portrait of Henry Ward Beecher* which stands upon the top of your books. You then glanced up at the wall, and of course your meaning was obvious. You were thinking that if the portrait were framed, it would just cover that bare space and correspond with Gordon's picture over there.'

'You have followed me wonderfully!' I exclaimed.

'So far I could hardly have gone astray. But now your thoughts went back to Beecher, and you looked hard across as if you were studying the character in his features. Then your eyes ceased to pucker, but you continued to look across, and your face was thoughtful. You were recalling the incidents of Beecher's career. I was well aware that you could not do this without thinking of the mission which he undertook on behalf of the North at the time of the Civil War,* for I remember your expressing your passionate indignation at the way in which he was received by the more turbulent of our people. You felt so strongly about it, that I knew you could not think of Beecher without thinking of that also. When a moment later I saw your eyes wander away from the picture, I suspected that your mind had now turned to the Civil War, and when I observed that your lips set, your eyes sparkled, and your hands clenched, I was positive that you were indeed thinking of the gallantry which was shown by both sides in that desperate struggle. But then, again, your face grew sadder; your shook your head. You were dwelling upon the sadness and horror and useless waste of life. Your hand stole towards your own old wound and a smile quivered on your lips, which showed me that the ridiculous side of this method of settling international questions had forced itself upon your mind. At this point I agreed with you that it was preposterous, and was glad to find that all my deductions had been correct.'

'Absolutely!' said I. 'And now that you have explained it, I confess that I am as amazed as before.'

'It was very superficial, my dear Watson, I assure you. I should not have intruded it upon your attention had you not shown some incredulity the other day. But I have in my hands here a little problem which may prove to be more difficult of solution than my small essay in thought reading. Have you observed in the paper a short paragraph referring to the remarkable contents of a packet sent through the post to Miss Susan Cushing, of Cross Street, Croydon?'

'No, I saw nothing.'

'Ah! then you must have overlooked it. Just toss it over to me. Here it is, under the financial column. Perhaps you would be good enough to read it aloud.'

I picked up the paper which he had thrown back to me, and read the paragraph indicated. It was headed, 'A Gruesome Packet.'

'Miss Susan Cushing, living at Cross Street, Croydon, has been made the victim of what must be regarded as a peculiarly revolting practical joke, unless some more sinister meaning should prove to be attached to the incident. At two o'clock yesterday afternoon a small packet, wrapped in brown paper, was handed in by the postman. A cardboard box was inside, which was filled with coarse salt. On emptying this, Miss Cushing was horrified to find two human ears, apparently quite freshly severed. The box had been sent by parcel post from Belfast upon the morning before. There is no indication as to the sender, and the matter is the more mysterious as Miss Cushing, who is a maiden lady of fifty, has led a most retired life, and has so few acquaintances or correspondents that it is a rare event for her to receive anything through the post. Some years ago, however, when she resided at Penge, she let apartments in her house to three young medical students, whom she was obliged to get rid of on account of their noisy and irregular habits. The police are of opinion that this outrage may have been perpetrated upon Miss Cushing by these youths, who owed her a grudge, and who hoped to frighten her by sending her these relics of the dissecting-rooms. Some probability is lent to the theory by the fact that one of these students came from the north of Ireland, and, to the best of Miss Cushing's belief, from Belfast. In the meantime, the matter is being actively investigated, Mr Lestrade, one of the very smartest of our detective officers, being in charge of the case.'

'So much for the *Daily Chronicle*,'* said Holmes, as I finished reading. 'Now for our friend Lestrade. I had a note from him this morning, in which he says: "I think that this case is very much in your line. We have every hope of clearing the matter up, but we find a little difficulty in getting anything to work upon. We have, of course, wired to the Belfast post-office, but a large number of parcels were handed in upon that day, and they have no means of identifying this particular one, or of remembering the sender. The box is a half-pound box of honeydew tobacco,* and does not help us in any way. The medical student theory still appears to me to be the most feasible, but if you should have a few hours to spare, I should be very happy to see you out here. I shall be either at the house or in the police-station all day." What say you, Watson? Can you rise superior to the heat, and run down to Croydon with me on the off chance of a case for your annals?'

'I was longing for something to do.'

'You shall have it, then. Ring for our boots, and tell them to order a cab. I'll be back in a moment, when I have changed my dressing-gown and filled my cigar-case.'

A shower of rain fell while we were in the train, and the heat was far less oppressive in Croydon than in town. Holmes had sent on a wire, so that Lestrade, as wiry, as dapper, and as ferret-like as ever, was waiting for us at the station. A walk of five minutes took us to Cross Street, where Miss Cushing resided.

It was a very long street of two-story brick houses, neat and prim, with whitened stone steps and little groups of aproned women gossiping at the doors. Half-way down, Lestrade stopped and tapped at a door, which was opened by a small servant girl. Miss Cushing was sitting in the front room, into which we were ushered. She was a placid-faced woman with large, gentle eyes, and grizzled hair curving down over her temples on each side. A worked antimacassar* lay upon her lap and a basket of coloured silks stood upon a stool beside her.

'They are in the outhouse, those dreadful things,' said she, as Lestrade entered. 'I wish that you would take them away altogether.'

'So I shall, Miss Cushing. I only kept them here until my friend, Mr Holmes, should have seen them in your presence.'

'Why in my presence, sir?'

'In case he wished to ask any questions.'

'What is the use of asking me questions, when I tell you that I know nothing whatever about it?'

'Quite so, madam,' said Holmes, in his soothing way. 'I have no doubt that you have been annoyed more than enough already over this business.'

'Indeed, I have, sir. I am a quiet woman and live a retired life. It is something new for me to see my name in the papers and to find the police in my house. I won't have those things in here, Mr Lestrade. If you wish to see them you must go to the outhouse.'

It was a small shed in the narrow garden which ran down behind the house. Lestrade went in and brought out a yellow cardboard box, with a piece of brown paper and some string. There was a bench at the edge of the path, and we all sat down while Holmes examined, one by one, the articles which Lestrade had handed to him.

'The string is exceedingly interesting,' he remarked, holding it up to the light and sniffing at it. 'What do you make of this string, Lestrade?'

'It has been tarred.'

35

'Precisely. It is a piece of tarred twine. You have also, no doubt, remarked that Miss Cushing has cut the cord with a scissors, as can be seen by the double fray on each side. This is of importance.'

'I cannot see the importance,' said Lestrade.

'The importance lies in the fact that the knot is left intact, and that this knot is of a peculiar character.'

'It is very neatly tied. I had already made a note to that effect,' said Lestrade, complacently.

'So much for the string then,' said Holmes, smiling; 'now for the box wrapper. Brown paper, with a distinct smell of coffee. What, you did not observe it? I think there can be no doubt of it. Address printed in rather straggling characters: "Miss S. Cushing, Cross Street, Croydon." Done with a broad pointed pen, probably a J,* and with very inferior ink. The word Croydon has been spelt originally with an i, which has been changed to y. The parcel was directed, then, by a man—the printing is distinctly masculine—of limited education and unacquainted with the town of Croydon. So far, so good! The box is a yellow, half-pound honeydew box, with nothing distinctive save two thumb marks at the left bottom corner. It is filled with rough salt of the quality used for preserving hides and other of the coarser commercial purposes. And embedded in it are these very singular inclosures.'

He took out the two ears as he spoke, and laying a board across his knees, he examined them minutely, while Lestrade and I, bending forward on each side of him, glanced alternately at these dreadful relics and at the thoughtful, eager face of our companion. Finally he returned them to the box once more, and sat for a while in deep thought.*

'You have observed, of course,' said he at last, 'that the ears are not a pair.'

'Yes, I have noticed that. But if this were the practical joke of some students from the dissecting-rooms, it would be as easy for them to send two odd ears as a pair.'

'Precisely. But this is not a practical joke.'

'You are sure of it?'

'The presumption is strongly against it. Bodies in the dissecting-rooms are injected with preservative fluid. These ears bear no signs of this. They are fresh, too. They have been cut off with a blunt instrument, which would hardly happen if a student had done it. Again, carbolic or rectified spirits* would be the preservatives which would suggest themselves to the medical mind, certainly not rough salt. I repeat that there is no practical joke here, but that we are investigating a serious crime.'

A vague thrill ran through me as I listened to my companion's words and saw the stern gravity which had hardened his features. This brutal preliminary seemed to shadow forth some strange and inexplicable horror in the background. Lestrade, however, shook his head like a man who is only half convinced.

'There are objections to the joke theory, no doubt,' said he; 'but there are much stronger reasons against the other. We know that this woman has led a most quiet and respectable life at Penge and here for the last twenty years. She has hardly been away from her home for a day during that time. Why on earth, then, should any criminal send her the proofs of his guilt, especially as, unless she is a most consummate actress, she understands quite as little of the matter as we do?'

'That is the problem which we have to solve,' Holmes answered, 'and for my part I shall set about it by presuming that my reasoning is correct, and that a double murder has been committed. One of these ears is a woman's, small, finely formed, and pierced for an earring. The other is a man's, sunburned, discoloured, and also pierced for an earring. These two people are presumably dead, or we should have heard their story before now. To-day is Friday. The packet was posted on Thursday morning. The tragedy, then, occurred on Wednesday or Tuesday, or earlier. If the two people were murdered, who but their murderer would have sent this sign of his work to Miss Cushing? We may take it that the sender of the packet is the man whom we want. But he must have some strong reason for sending Miss

Cushing this packet. What reason, then? It must have been to tell her that the deed was done; or to pain her, perhaps. But in that case she knows who it is. Does she know? I doubt it. If she knew, why should she call the police in? She might have buried the ears, and no one would have been the wiser. That is what she would have done if she had wished to shield the criminal. But if she does not wish to shield him she would give his name. There is a tangle here which needs straightening out.' He had been talking in a high, quick voice, staring blankly up over the garden fence, but now he sprang briskly to his feet and walked towards the house.

'I have a few questions to ask Miss Cushing,' said he.

'In that case I may leave you here,' said Lestrade, 'for I have another small business on hand. I think that I have nothing further to learn from Miss Cushing. You will find me at the police-station.'

'We shall look in on our way to the train,' answered Holmes. A moment later he and I were back in the front room, where the impassive lady was still quietly working away at her antimacassar. She put it down on her lap as we entered, and looked at us with her frank, searching blue eyes.

'I am convinced, sir,' she said, 'that this matter is a mistake, and that the parcel was never meant for me at all. I have said this several times to the gentleman from Scotland Yard, but he simply laughs at me. I have not an enemy in the world, as far as I know, so why should anyone play me such a trick?'

'I am coming to be of the same opinion, Miss Cushing,' said Holmes, taking a seat beside her. 'I think that it is more than probable——' he paused, and I was surprised on glancing round to see that he was staring with singular intentness at the lady's profile. Surprise and satisfaction were both for an instant to be read upon his eager face, though when she glanced round to find out the cause of his silence he had become as demure as ever. I stared hard myself at her flat, grizzled hair, her trim cap, her little gilt earrings, her placid features; but I could see nothing which could account for my companion's evident excitement.

'There were one or two questions——'

'Oh, I am weary of questions!' cried Miss Cushing, impatiently.

'You have two sisters, I believe.'

'How could you know that?'

'I observed the very instant that I entered the room that you have a portrait group of three ladies upon the mantelpiece, one of whom is undoubtedly yourself, while the others are so exceedingly like you that there could be no doubt of the relationship.'

'Yes, you are quite right. Those are my sisters, Sarah and Mary.'

'And here at my elbow is another portrait, taken at Liverpool, of your younger sister, in the company of a man who appears to be a steward by his uniform. I observe that she was unmarried at the time.'

'You are very quick at observing.'

'That is my trade.'

'Well, you are quite right. But she was married to Mr Browner a few days afterwards. He was on the South American line when that was taken, but he was so fond of her that he couldn't abide to leave her for so long, and he got into the Liverpool and London boats.'

'Ah, the *Conqueror*, perhaps?'

'No, the *May Day*,* when last I heard. Jim came down here to see me once. That was before he broke the pledge;* but afterwards he would always take drink when he was ashore, and a little drink would send him stark, staring mad. Ah! it was a bad day that ever he took a glass in his hand again. First he dropped me, and then he quarrelled with Sarah, and now that Mary has stopped writing we don't know how things are going with them.'

It was evident that Miss Cushing had come upon a subject on which she felt very deeply. Like most people who lead a lonely life, she was shy at first, but ended by becoming extremely communicative. She told us many details about her brother-in-law the steward, and then wandering off on to the subject of her former lodgers, the medical students, she gave us a long account of their delinquencies, with their names and those of their hospitals. Holmes listened attentively to everything, throwing in a question from time to time.

'About your second sister, Sarah,' said he. 'I wonder, since you are both maiden ladies, that you do not keep house together.'

'Ah! you don't know Sarah's temper, or you would wonder no more. I tried it when I came to Croydon, and we kept on until about two months ago, when we had to part. I don't want to say a word against my own sister, but she was always meddlesome and hard to please, was Sarah.'

'You say that she quarrelled with your Liverpool relations.'

'Yes, and they were the best of friends at one time. Why, she went up there to live just in order to be near them. And now she has no word hard enough for Jim Browner. The last six months that she was here she would speak of nothing but his drinking and his ways. He had caught her meddling, I suspect, and given her a bit of his mind, and that was the start of it.'

'Thank you, Miss Cushing,' said Holmes, rising and bowing. 'Your sister Sarah lives, I think you said, at New Street, Wallington? Good-bye, and I am very sorry that you should have been troubled over a case with which, as you say, you have nothing whatever to do.'

There was a cab passing as we came out, and Holmes hailed it.

'How far to Wallington?' he asked.

'Only about a mile, sir.'

'Very good. Jump in, Watson. We must strike while the iron is hot. Simple as the case is, there have been one or two very instructive details in connection with it. Just pull up at a telegraph office as you pass, cabby.'

Holmes sent off a short wire, and for the rest of the drive lay back in the cab with his hat tilted over his nose to keep the sun from his face. Our driver pulled up at a house which was not unlike the one which we had just quitted. My companion ordered him to wait, and had his hand upon the knocker, when the door opened and a grave young gentleman in black, with a very shiny hat, appeared on the step.

'Is Miss Sarah Cushing at home?' asked Holmes.

'Miss Sarah Cushing is extremely ill,' said he. 'She has been suffering since yesterday from brain symptoms of great severity. As her medical adviser, I cannot possibly take the responsibility of allowing anyone to see her. I should recommend you to call again in ten days.' He drew on his gloves, closed the door, and marched off down the street.

'Well, if we can't, we can't,' said Holmes, cheerfully.

'Perhaps she could not, or would not have told you much.'

'I did not wish her to tell me anything. I only wanted to look at her. However, I think that I have got all that I want. Drive us to some decent hotel, cabby, where we may have some lunch, and afterwards we shall drop down upon friend Lestrade at the police-station.'

We had a pleasant little meal together, during which Holmes would talk about nothing but violins, narrating with great exultation how he had purchased his own Stradivarius,* which was worth at least five hundred guineas, at a Jew broker's* in Tottenham Court Road for fifty-five shillings. This led him to Paganini,* and we sat for an hour over a bottle of claret* while he told me anecdote after anecdote of that extraordinary man. The afternoon was far advanced and the hot glare had softened into a mellow glow before we found ourselves at the police-station. Lestrade was waiting for us at the door.

'A telegram for you, Mr Holmes,' said he.

'Ha! It is the answer!' He tore it open, glanced his eyes over it, and crumpled it into his pocket. 'That's all right,' said he.

'Have you found out anything?'

'I have found out everything!'

'What!' Lestrade stared at him in amazement. 'You are joking.'

'I was never more serious in my life. A shocking crime has been committed, and I think that I have now laid bare every detail of it.'

'And the criminal?'

Holmes scribbled a few words upon the back of one of his visiting cards and threw it over to Lestrade.

'That is it,'* he said; 'you cannot effect an arrest until to-morrow night at the earliest. I should prefer that you would not mention my name at all in connection with the case, as I choose to be associated only with those crimes which present some difficulty in their solution. Come on, Watson.' We strode off together to the station, leaving Lestrade still staring with a delighted face at the card which Holmes had thrown him.

'The case,' said Sherlock Holmes, as we chatted over our cigars that night in our rooms at Baker Street, 'is one where, as in the investigations which you have chronicled under the names of the 'Study in Scarlet' and of the 'Sign of Four,' we have been compelled to reason backward from effects to causes.* I have written to Lestrade asking him to supply us with the details which are now wanting, and which he will only get after he has secured his man. That he may be safely trusted to do, for although he is absolutely devoid of reason, he is as tenacious as a bull-dog when he once understands what he has to do, and indeed it is just this tenacity which has brought him to the top at Scotland Yard.'

'Your case is not complete, then?' I asked.

'It is fairly complete in essentials. We know who the author of the revolting business is, although one of the victims still escapes us. Of course, you have formed your own conclusions.'

'I presume that this Jim Browner, the steward of a Liverpool boat, is the man whom you suspect?'

'Oh! it is more than a suspicion.'

'And yet I cannot see anything save very vague indications.'

'On the contrary, to my mind nothing could be more clear. Let me run over the principal steps. We approached the case, you remember, with an absolutely blank mind, which is always an advantage. We had formed no theories.

We were simply there to observe and to draw inferences from our observations. What did we see first? A very placid and respectable lady, who seemed quite innocent of any secret, and a portrait which showed me that she had two younger sisters. It instantly flashed across my mind that the box might have been meant for one of these. I set the idea aside as one which could be disproved or confirmed at our leisure. Then we went to the garden, as you remember, and we saw the very singular contents of the little yellow box.

'The string was of the quality which is used by sail-makers aboard ship, and at once a whiff of the sea was perceptible in our investigation. When I observed that the knot was one which is popular with sailors, that the parcel had been posted at a port, and that the male ear was pierced for an earring which is so much more common among sailors than landsmen, I was quite certain that all the actors in the tragedy were to be found among our seafaring classes.

'When I came to examine the address of the packet I observed that it was to Miss S. Cushing. Now, the oldest sister would, of course, be Miss Cushing, and although her initial was "S.," it might belong to one of the others as well. In that case we should have to commence our investigation from a fresh basis altogether. I therefore went into the house with the intention of clearing up this point. I was about to assure Miss Cushing that I was convinced that a mistake had been made, when you may remember that I came suddenly to a stop. The fact was that I had just seen something which filled me with surprise, and at the same time narrowed the field of our inquiry immensely.

'As a medical man, you are aware, Watson, that there is no part of the body which varies so much as the human ear. Each ear is as a rule quite distinctive, and differs from all other ones. In last year's *Anthropological Journal** you will find two short monographs from my pen upon the subject. I had therefore examined the ears in the box with the eyes of an expert, and had carefully noted their anatomical peculiarities. Imagine my surprise then, when, on looking at Miss

Cushing, I perceived that her ear corresponded exactly with the female ear which I had just inspected. The matter was entirely beyond coincidence. There was the same shortening of the pinna,* the same broad curve of the upper lobe, the same convolution of the inner cartilage. In all essentials it was the same ear.

'Of course, I at once saw the enormous importance of the observation. It was evident that the victim was a blood relation, and probably a very close one. I began to talk to her about her family, and you remember that she at once gave us some exceedingly valuable details.

'In the first place, her sister's name was Sarah, and her address had, until recently, been the same, so that it was quite obvious how the mistake had occurred, and whom the packet was meant for. Then we heard of this steward, married to the third sister, and learned that he had at one time been so intimate with Miss Sarah that she had actually gone up to Liverpool to be near the Browners, but a quarrel had afterwards divided them. This quarrel had put a stop to all communications for some months, so that if Browner had occasion to address a packet to Miss Sarah, he would undoubtedly have done so to her old address.

'And now the matter had begun to straighten itself out wonderfully. We had learned of the existence of this steward, an impulsive man, of strong passions—you remember that he threw up what must have been a very superior berth, in order to be nearer to his wife—subject, too, to occasional fits of hard drinking. We had reason to believe that his wife had been murdered, and that a man—presumably a seafaring man—had been murdered at the same time. Jealousy, of course, at once suggests itself as the motive for the crime. And why should these proofs of the deed be sent to Miss Sarah Cushing? Probably because during her residence in Liverpool she had some hand in bringing about the events which led to the tragedy. You will observe that this line of boats calls at Belfast, Dublin, and Waterford; so that, presuming that Browner had committed the deed, and had embarked at once upon his steamer, the *May Day*, Belfast would be the first place at which he could post his terrible packet.

'A second solution was at this stage obviously possible, and although I thought it exceedingly unlikely, I was determined to elucidate it before going further. An unsuccessful lover might have killed Mr and Mrs Browner, and the male ear might have belonged to the husband. There were many grave objections to this theory, but it was conceivable. I therefore sent off a telegram to my friend Algar, of the Liverpool force, and asked him to find out if Mrs Browner were at home, and if Browner had departed in the *May Day*. Then we went on to Wallington to visit Miss Sarah.

'I was curious, in the first place, to see how far the family ear had been reproduced in her. Then, of course, she might give us very important information, but I was not sanguine that she would. She must have heard of the business the day before, since all Croydon was ringing with it, and she alone could have understood whom the packet was meant for. If she had been willing to help justice she would probably have communicated with the police already. However, it was clearly our duty to see her, so we went. We found that the news of the arrival of the packet—for her illness dated from that time—had such an effect upon her as to bring on brain fever. It was clearer than ever that she understood its full significance, but equally clear that we should have to wait some time for any assistance from her.

'However, we were really independent of her help. Our answers were waiting for us at the police-station, where I had directed Algar to send them. Nothing could be more conclusive. Mrs Browner's house had been closed for more than three days, and the neighbours were of opinion that she had gone south to see her relatives. It had been ascertained at the shipping offices that Browner had left aboard of the *May Day*, and I calculate that she is due in the Thames tomorrow night. When he arrives he will be met by the obtuse but resolute Lestrade, and I have no doubt that we shall have all our details filled in.'

*

Sherlock Holmes was not disappointed in his expectations. Two days later he received a bulky envelope, which contained a short note from the detective, and a type-written document, which covered several pages of foolscap.

'Lestrade has got him all right,' said Holmes, glancing up at me. 'Perhaps it would interest you to hear what he says.'

'My dear Mr. Holmes,—In accordance with the scheme which we had formed in order to test our theories'—'the "we" is rather fine, Watson, is it not?'—'I went down to the Albert Dock* yesterday at 6 p.m., and boarded the ss. *May Day*, belonging to the Liverpool, Dublin, and London Steam Packet Company. On inquiry, I found that there was a steward on board of the name of James Browner, and that he had acted during the voyage in such an extraordinary manner that the captain had been compelled to relieve him of his duties. On descending to his berth, I found him seated upon a chest with his head sunk upon his hands, rocking himself to and fro. He is a big, powerful chap, clean-shaven, and very swarthy—something like Aldridge, who helped us in the bogus laundry affair. He jumped up when he heard my business, and I had my whistle to my lips to call a couple of river police, who were round the corner, but he seemed to have no heart in him, and he held out his hands quietly enough for the darbies.* We brought him along to the cells, and his box as well, for we thought there might be something incriminating; but, bar a big sharp knife, such as most sailors have, we got nothing for our trouble. However, we find that we shall want no more evidence, for, on being brought before the inspector at the station, he asked leave to make a statement, which was, of course, taken down, just as he made it, by our shorthand man. We had three copies type-written, one of which I inclose. The affair proves, as I always thought it would, to be an extremely simple one, but I am obliged to you for assisting me in my investigation. With kind regards, yours very truly,—G. LESTRADE.'

*

'Hum! The investigation really was a very simple one,' remarked Holmes; 'but I don't think it struck him in that light when he first called us in. However, let us see what Jim Browner has to say for himself. This is his statement, as made before Inspector Montgomery at the Shadwell* Police Station, and it has the advantage of being verbatim.'

'Have I anything to say? Yes, I have a deal to say. I have to make a clean breast of it all. You can hang me, or you can leave me alone. I don't care a plug* which you do. I tell you I've not shut an eye in sleep since I did it, and I don't believe I ever will again until I get past all waking. Sometimes it's his face, but most generally it's hers. I'm never without one or the other before me. He looks frowning and black-like, but she has a kind o' surprise upon her face. Aye, the white lamb, she might well be surprised when she read death on a face that had seldom looked anything but love upon her before.

'But it was Sarah's fault, and may the curse of a broken man put a blight on her and set the blood rotting in her veins! It's not that I want to clear myself. I know that I went back to drink, like the beast that I was. But she would have forgiven me; she would have stuck as close to me as a rope to a block if that woman had never darkened our door. For Sarah Cushing loved me—that's the root of the business— she loved me, until all her love turned to poisonous hate when she knew that I thought more of my wife's footmark in the mud than I did of her whole body and soul.

'There were three sisters altogether. The old one was just a good woman, the second was a devil, and the third was an angel. Sarah was thirty-three, and Mary was twenty-nine when I married. We were just as happy as the day was long when we set up house together, and in all Liverpool there was no better woman than my Mary. And then we asked Sarah up for a week, and the week grew into a month, and one thing led to another, until she was just one of ourselves.

'I was blue ribbon* at that time, and we were putting a little money by, and all was as bright as a new dollar. My

God, whoever would have thought that it could have come to this? Whoever would have dreamed it?

'I used to be home for the week-ends very often, and sometimes if the ship were held back for cargo I would have a whole week at a time, and in this way I saw a deal of my sister-in-law, Sarah. She was a fine tall woman, black and quick and fierce, with a proud way of carrying her head, and a glint from her eye like the spark from a flint. But when little Mary was there I had never a thought for her, and that I swear as I hope for God's mercy.

'It had seemed to me sometimes that she liked to be alone with me, or to coax me out for a walk with her, but I had never thought anything of that. But one evening my eyes were opened. I had come up from the ship and found my wife out, but Sarah at home. "Where's Mary?" I asked. "Oh, she has gone to pay some accounts." I was impatient and paced up and down the room. "Can't you be happy for five minutes without Mary, Jim?" says she. "It's a bad compliment to me that you can't be contented with my society for so short a time." "That's all right, my lass," said I, putting out my hand towards her in a kindly way, but she had it in both hers in an instant, and they burned as if they were in a fever. I looked into her eyes and I read it all there. There was no need\for her to speak, nor for me either. I frowned and drew my hand away. Then she stood by my side in silence for a bit, and then put up her hand and patted me on the shoulder. "Steady old Jim!" said she; and, with a kind o' mocking laugh, she ran out of the room.

'Well, from that time Sarah hated me with her whole heart and soul, and she is a woman who can hate, too. I was a fool to let her go on biding with us—a besotted fool—but I never said a word to Mary, for I knew it would grieve her. Things went on much as before, but after a time I began to find that there was a bit of a change in Mary herself. She had always been so trusting and so innocent, but now she became queer and suspicious, wanting to know where I had been and what I had been doing, and whom my letters were from, and what I had in my pockets, and a thousand such follies. Day by day she grew queerer and more irritable, and we had causeless rows about nothing. I was fairly puzzled by it all. Sarah avoided me now, but she and Mary were just inseparable. I can see now how she was plotting and scheming and poisoning my wife's mind against me, but I was such a blind beetle that I could not understand it at the time. Then I broke my blue ribbon and began to drink again, but I think I should not have done it if Mary had been the same as ever. She had some reason to be disgusted with me now, and the gap between us began to be wider and wider. And then this Alec Fairbairn chipped in, and things became a thousand times blacker.

'It was to see Sarah that he came to my house first, but soon it was to see us, for he was a man with winning ways, and he made friends wherever he went. He was a dashing, swaggering chap, smart and curled, who had seen half the world, and could talk of what he had seen. He was good company, I won't deny it, and he had wonderful polite ways with him for a sailor man, so that I think there must have been a time when he knew more of the poop than the forecastle.* For a month he was in and out of my house, and never once did it cross my mind that harm might come of his soft, tricky ways. And then at last something made me suspect, and from that day my peace was gone for ever.

'It was only a little thing, too. I had come into the parlour unexpected, and as I walked in at the door I saw a light of welcome on my wife's face. But as she saw who it was it faded again, and she turned away with a look of disappointment. That was enough for me. There was no one but Alec Fairbairn whose step she could have mistaken for mine. If I could have seen him then I should have killed him, for I have always been like a madman when my temper gets loose. Mary saw the devil's light in my eyes, and she ran forward with her hands on my sleeve. "Don't, Jim, don't!" says she. "Where's Sarah?" I asked. "In the kitchen," says she. "Sarah," says I, as I went in, "this man Fairbairn is never to darken my door again." "Why not?" says she. "Because I order it." "Oh!" says she, "if my friends are not

49

good enough for this house, then I am not good enough for it either." "You can do what you like," says I, "but if Fairbairn shows his face here again, I'll send you one of his ears for a keepsake." She was frightened by my face, I think, for she never answered a word, and the same evening she left my house.

'Well, I don't know now whether it was pure devilry on the part of this woman, or whether she thought that she could turn me against my wife by encouraging her to misbehave. Anyway, she took a house just two streets off, and let lodgings to sailors. Fairbairn used to stay there, and Mary would go round to have tea with her sister and him. How often she went I don't know, but I followed her one day, and as I broke in at the door Fairbairn got away over the back garden wall, like the cowardly skunk that he was. I swore to my wife that I would kill her if I found her in his company again, and I led her back with me, sobbing and trembling, and as white as a piece of paper. There was no trace of love between us any longer. I could see that she hated me and feared me, and when the thought of it drove me to drink, then she despised me as well.

'Well, Sarah found that she could not make a living in Liverpool, so she went back, as I understand, to live with her sister in Croydon, and things jogged on much the same as ever at home. And then came this last week and all the misery and ruin.

'It was in this way. We had gone on the *May Day* for a round voyage of seven days, but a hogshead* got loose and started one of our plates,* so that we had to put back into port for twelve hours. I left the ship and came home, thinking what a surprise it would be for my wife, and hoping that maybe she would be glad to see me so soon. The thought was in my head as I turned into my own street, and at that moment a cab passed me, and there she was, sitting by the side of Fairbairn, the two chatting and laughing, with never a thought for me as I stood watching them from the footpath.

'I tell you, and I give you my word on it, that from that moment I was not my own master, and it is all like a dim dream when I look back on it. I had been drinking hard of late, and the two things together fairly turned my brain. There's something throbbing in my head now, like a docker's hammer, but that morning I seemed to have all Niagara whizzing and buzzing in my ears.

'Well, I took to my heels, and I ran after the cab. I had a heavy oak stick in my hand, and I tell you that I saw red from the first; but as I ran I got cunning, too, and hung back a little to see them without being seen. They pulled up soon at the railway station. There was a good crowd round the booking-office, so I got quite close to them without being seen. They took tickets for New Brighton.* So did I, but I got in three carriages behind them. When we reached it they walked along the Parade, and I was never more than a hundred yards from them. At last I saw them hire a boat and start for a row, for it was a very hot day, and they thought no doubt that it would be cooler on the water.

'It was just as if they had been given into my hands. There was a bit of a haze, and you could not see more than a few hundred yards. I hired a boat for myself, and I pulled after them. I could see the blurr of their craft, but they were going nearly as fast as I, and they must have been a long mile from the shore before I caught them up. The haze was like a curtain all round us, and there were we three in the middle of it. My God, shall I ever forget their faces when they saw who was in the boat that was closing in upon them? She screamed out. He swore like a madman, and jabbed at me with an oar, for he must have seen death in my eyes. I got past it and got one in with my stick, that crushed his head like an egg. I would have spared her, perhaps, for all my madness, but she threw her arms round him, crying out to him, and calling him "Alec." I struck again, and she lay stretched beside him. I was like a wild beast then that had tasted blood. If Sarah had been there, by the Lord, she should have joined them. I pulled out my knife, and—well, there! I've said enough. It gave me a kind of savage joy when I thought how Sarah would feel when she had such signs as these of what her meddling had brought about. Then I tied the bodies into the boat, stove a plank, and stood by until they had sunk. I knew very well that the owner would think that they had lost their bearings in the haze, and had drifted off out to sea. I cleaned myself up, got back to land, and joined my ship without a soul having a suspicion of what had passed. That night I made up the packet for Sarah Cushing, and next day I sent it from Belfast.

'There you have the whole truth of it. You can hang me, or do what you like with me, but you cannot punish me as I have been punished already. I cannot shut my eyes but I see those two faces staring at me—staring at me as they stared when my boat broke through the haze. I killed them quick, but they are killing me slow; and if I have another night of it I shall be either mad or dead before morning. You won't put me alone into a cell, sir? For pity's sake don't, and may you be treated in your day of agony as you treat me now.'

'What is the meaning of it, Watson?' said Holmes, solemnly, as he laid down the paper. 'What object is served by this circle of misery and violence and fear? It must tend to some end, or else our universe is ruled by chance, which is unthinkable. But what end? There is the great standing perennial problem to which human reason is as far from an answer as ever.'

The Yellow Face

IN publishing these short sketches, based upon the numerous cases in* which my companion's singular gifts have made me the listener to, and eventually the actor in some strange drama, it is only natural that I should dwell rather upon his successes than upon his failures. And this is not so much for the sake of his reputation, for indeed it was when he was at his wits' end that his energy and his versatility were most admirable, but because where he failed it happened too often that no one else succeeded, and that the tale was left for ever without a conclusion. Now and again, however, it chanced that even when he erred the truth was still discovered. I have notes of some half-dozen cases of the kind, of which the affair of the second stain,* and that which I am now about to recount, are the two which present the strongest features of interest.

Sherlock Holmes was a man who seldom took exercise for exercise's sake. Few men were capable of greater muscular effort, and he was undoubtedly one of the finest boxers of his weight that I have ever seen;* but he looked upon aimless bodily exertion as a waste of energy, and he seldom bestirred himself save where there was some professional object to be served. Then he was absolutely untiring and indefatigable. That he should have kept himself in training under such circumstances is remarkable, but his diet was usually of the sparest, and his habits were simple to the verge of austerity. Save for the occasional use of cocaine* he had no vices, and he only turned to the drug as a protest against the monotony of existence when cases were scanty and the papers uninteresting.

One day in early spring he had so far relaxed as to go for a walk with me in the Park,* where the first faint shoots of green were breaking out upon the elms, and the sticky spearheads of the chestnuts were just beginning to burst into their five-fold leaves. For two hours we rambled about together, in silence for the most part, as befits two men who know each other intimately. It was nearly five before we were back in Baker Street once more.

'Beg pardon, sir,' said our page-boy,* as he opened the door; 'there's been a gentleman here asking for you, sir.'

Holmes glanced reproachfully at me. 'So much for afternoon walks!' said he. 'Has this gentleman gone, then?'

'Yes, sir.'

'Didn't you ask him in?'

'Yes, sir; he came in.'

'How long did he wait?'

'Half an hour, sir. He was a very restless gentleman, sir, a walkin' and a stampin' all the time he was here. I was waitin' outside the door, sir, and I could hear him. At last he goes out into the passage and he cries: "Is that man never goin' to come?" Those were his very words, sir. "You'll only need to wait a little longer," says I. "Then I'll wait in the open air, for I feel half choked," says he. "I'll be back before long," and with that he ups and he outs, and all I could say wouldn't hold him back.'

'Well, well, you did your best,' said Holmes, as we walked into our room. 'It's very annoying though, Watson. I was badly in need of a case, and this looks, from the man's impatience, as if it were of importance. Hullo! that's not your pipe on the table! He must have left his behind him. A nice old briar, with a good long stem of what the tobacconists call amber.* I wonder how many real amber mouthpieces there are in London. Some people think a fly in it is a sign. Why, it is quite a branch of trade, the putting of sham flies into the sham amber.* Well, he must have been disturbed in his mind to leave a pipe behind him which he evidently values highly.'

'How do you know that he values it highly?' I asked.

'Well, I should put the original cost of the pipe at seven-and-sixpence.* Now it has, you see, been twice mended: once in the wooden stem and once in the amber. Each of these mends, done, as you observe, with silver bands, must have cost more than the pipe did originally. The man must value the pipe highly when he prefers to patch it up rather than buy a new one with the same money.'

'Anything else?' I asked, for Holmes was turning the pipe about in his hand and staring at it in his peculiar, pensive way.

He held it up and tapped on it with his long, thin fore-finger as a professor might who was lecturing on a bone.

'Pipes are occasionally of extraordinary interest,' said he. 'Nothing has more individuality save, perhaps, watches and bootlaces. The indications here, however, are neither very marked nor very important. The owner is obviously a muscular man, left-handed, with an excellent set of teeth, careless in his habits, and with no need to practise economy.'

My friend threw out the information in a very off-hand way, but I saw that he cocked his eye at me to see if I had followed his reasoning.

'You think a man must be well-to-do if he smokes a seven-shilling pipe?' said I.

'This is Grosvenor mixture at eightpence* an ounce,' Holmes answered, knocking a little out on his palm. 'As he might get an excellent smoke for half the price, he has no need to practise economy.'

'And the other points?'

'He has been in the habit of lighting his pipe at lamps and gas-jets. You can see that it is quite charred all down one side. Of course, a match could not have done that. Why should a man hold a match to the side of his pipe? But you cannot light it at a lamp without getting the bowl charred. And it is all on the right side of the pipe. From that I gather that he is a left-handed man. You hold your own pipe to the lamp, and see how naturally you, being right-handed, hold the left side to the flame. You might do it once the other way, but not as a constancy. This has always been held so. Then he has bitten through his amber. It takes a muscular, energetic fellow, and one with a good set of teeth to do that.

But if I am not mistaken I hear him upon the stair, so we shall have something more interesting than his pipe to study.'

An instant later our door opened, and a tall young man entered the room. He was well but quietly dressed in a dark-grey suit, and carried a brown wide-awake* in his hand. I should have put him at about thirty, though he was really some years older.

'I beg your pardon,' said he, with some embarrassment; 'I suppose I should have knocked. Yes, of course I should have knocked. The fact is that I am a little upset, and you must put it all down to that.' He passed his hand over his forehead like a man who is half dazed, and then fell, rather than sat, down upon a chair.

'I can see that you have not slept for a night or two,' said Holmes, in his easy, genial way. 'That tries a man's nerves more than work, and more even than pleasure. May I ask how I can help you?'

'I wanted your advice, sir. I don't know what to do, and my whole life seems to have gone to pieces.'

'You wish to employ me as a consulting detective?'

'Not that only. I want your opinion as a judicious man— as a man of the world. I want to know what I ought to do next. I hope to God you'll be able to tell me.'

He spoke in little, sharp, jerky outbursts, and it seemed to me that to speak at all was very painful to him, and that his will all through was overriding his inclinations.

'It's a very delicate thing,' said he. 'One does not like to speak of one's domestic affairs to strangers. It seems dreadful to discuss the conduct of one's wife with two men whom I have never seen before. It's horrible to have to do it. But I've got to the end of my tether, and I must have advice.'

'My dear Mr Grant Munro*—' began Holmes.

Our visitor sprang from his chair. 'What!' he cried. 'You know my name?'

'If you wish to preserve your *incognito*,'* said Holmes, smiling, 'I should suggest that you cease to write your name upon the lining of your hat, or else that you turn the crown towards the person whom you are addressing. I was about to say that my friend and I have listened to many strange secrets in this room, and that we have had the good fortune to bring peace to many troubled souls. I trust that we may do as much for you. Might I beg you, as time may prove to be of importance, to furnish me with the facts of your case without further delay?'

Our visitor again passed his hand over his forehead as if he found it bitterly hard. From every gesture and expression I could see that he was a reserved, self-contained man, with a dash of pride in his nature, more likely to hide his wounds than to expose them. Then suddenly, with a fierce gesture of his closed hand, like one who throws reserve to the winds, he began.

'The facts are these, Mr Holmes,' said he. 'I am a married man, and have been so for three years. During that time my wife and I have loved each other as fondly, and lived as happily, as any two that ever were joined. We have not had a difference, not one, in thought, or word, or deed. And now, since last Monday, there has suddenly sprung up a barrier between us, and I find that there is something in her life and in her thoughts of which I know as little as if she were the woman who brushes by me in the street. We are estranged, and I want to know why.

'Now there is one thing I want to impress upon you before I go any further, Mr Holmes: Effie loves me. Don't let there be any mistake about that. She loves me with her whole heart and soul, and never more than now. I know it, I feel it. I don't want to argue about that. A man can tell easily enough when a woman loves him. But there's this secret between us, and we can never be the same until it is cleared.'

'Kindly let me have the facts, Mr Munro,' said Holmes, with some impatience.

'I'll tell you what I know about Effie's history. She was a widow when I met her first, though quite young—only twenty-five. Her name then was Mrs Hebron. She went out to America when she was young and lived in the town of

Atlanta,* where she married this Hebron, who was a lawyer with a good practice. They had one child, but the yellow fever broke out badly in the place, and both husband and child died of it. I have seen his death certificate. This sickened her of America, and she came back to live with a maiden aunt at Pinner, in Middlesex. I may mention that her husband had left her comfortably off, and that she had a capital of about four thousand five hundred pounds, which had been so well invested by him that it returned an average of 7 per cent. She had only been six months at Pinner when I met her; we fell in love with each other, and we married a few weeks afterwards.

'I am a hop merchant myself, and as I have an income of seven or eight hundred, we found ourselves comfortably off, and took a nice eighty-pound-a-year villa at Norbury. Our little place was very countrified, considering that it is so close to town. We had an inn and two houses a little above us, and a single cottage at the other side of the field which faces us, and except those there were no houses until you get half-way to the station. My business took me into town at certain seasons, but in summer I had less to do, and then in our country home my wife and I were just as happy as could be wished. I tell you that there never was a shadow between us until this accursed affair began.

'There's one thing I ought to tell you before I go further. When we married, my wife made over all her property to me—rather against my will, for I saw how awkward it would be if my business affairs went wrong. However, she would have it so, and it was done. Well, about six weeks ago she came to me.

' "Jack,*" said she, "when you took my money you said that if ever I wanted any I was to ask you for it."

' "Certainly," said I, "it's all your own."

' "Well," said she, "I want a hundred pounds."

'I was a bit staggered at this, for I had imagined it was simply a new dress or something of the kind that she was after.

' "What on earth for?" I asked.

' "Oh," said she, in her playful way, "you said that you were only my banker, and bakers never ask questions, you know."

' "If you really mean it, of course you shall have the money," said I.

' "Oh, yes, I really mean it."

' "And you won't tell me what you want it for?"

' "Some day, perhaps, but not just at present, Jack."

'So I had to be content with that, though it was the first time that there had ever been any secret between us. I gave her a cheque, and I never thought any more of the matter. It may have nothing to do with what came afterwards, but I thought it only right to mention it.

'Well, I told you just now that there is a cottage not far from our house. There is just a field between us, but to reach it you have to go along the road and then turn down a lane. Just beyond it is a nice little grove of Scotch firs, and I used to be very fond of strolling down there, for trees are always neighbourly kinds of things. The cottage had been standing empty this eight months, and it was a pity, for it was a pretty two-storied place, with an old-fashioned porch and honeysuckle about it. I have stood many a time and thought what a neat little homestead it would make.

'Well, last Monday evening I was taking a stroll down that way, when I met an empty van coming up the lane, and saw a pile of carpets and things lying about on te grass-plot beside the porch. It was clear that the cottage had at last been let. I walked past it, and then stopping, as an idle man might, I ran my eye over it, and wondered what sort of folk they were who had come to live so near us. And as I looked I suddenly became aware that a face was watching me out of one of the upper windows.

'I don't know what there was about that face, Mr Holmes, but it seemed to send a chill right down my back. I was some little way off, so that I could not make out the features, but there was something unnatural and inhuman about the face. That was the impression I had, and I moved quickly forwards to get a nearer view of the person who was watching me. But as I did so the face suddenly disappeared, so suddenly that it seemed to have been plucked away into the darkness of the room. I stood for five minutes thinking the business over, and trying to analyse my impressions. I could not tell if the face was that of a man or a woman. But the colour was what impressed me most. It was of a livid dead yellow, and with something set and rigid about it, which was shockingly unnatural. So disturbed was I, that I determined to see a little more of the new inmates of the cottage. I approached and knocked at the door, which was instantly opened by a tall, gaunt woman, with a harsh, forbidding face.

' "What may you be wantin'?" she asked, in a northern accent.

' "I am your neighbour over yonder," said I, nodding towards my house. "I see that you have only just moved in, so I thought that if I could be of any help to you in any—"

' "Aye, we'll just ask ye when we want ye," said she, and shut the door in my face. Annoyed at the churlish rebuff, I turned my back and walked home. All the evening, though I tried to think of other things, my mind would still turn to the apparition at the window and the rudeness of the woman. I determined to say nothing about the former to my wife, for she is a nervous, highly-strung woman, and I had no wish that she should share the unpleasant impression which had been produced upon myself. I remarked to her, however, before I fell asleep that the cottage was now occupied, to which she returned no reply.

'I am usually an extremely sound sleeper. It has been a standing jest in the family that nothing could ever wake me during the night; and yet somehow on that particular night, whether it may have been the slight excitement produced by my little adventure or not, I know not, but I slept much more lightly than usual. Half in my dreams I was dimly conscious that something was going on in the room, and gradually became aware that my wife had dressed herself and was slipping on her mantle and her bonnet. My lips were parted to murmur out some sleepy words of surprise or remonstrance at this untimely preparation, when suddenly my half-opened eyes fell upon her face, illuminated by the candle light, and astonishment held me dumb. She wore an expression such as I had never seen before—such as I should have thought her incapable of assuming. She was deadly pale, and breathing fast, glancing furtively towards the bed, as she fastened her mantle, to see if she had disturbed me. Then, thinking that I was still asleep, she slipped noiselessly from the room, and an instant later I heard a sharp creaking, which could only come from the hinges of the front door. I sat up in bed and rapped my knuckles against the rail to make certain that I was truly awake. Then I took my watch from under the pillow. It was three in the morning. What on earth could my wife be doing out on the country road at three in the morning?

'I had sat for about twenty minutes turning the thing over in my mind and trying to find some possible explanation. The more I thought, the more extraordinary and inexplicable did it appear. I was still puzzling over it when I heard the door gently close again and her footsteps coming up the stairs.

' "Where in the world have you been, Effie?" I asked, as she entered.

'She gave a violent start and a kind of gasping cry when I spoke, and that cry and start troubled me more than all the rest, for there was something indescribably guilty about them. My wife had always been a woman of a frank, open nature, and it gave me a chill to see her slinking into her own room, and crying out and wincing when her own husband spoke to her.

' "You awake, Jack?" she cried, with a nervous laugh. "Why, I thought that nothing could awaken you."

' "Where have you been?" I asked, more sternly.

' "I don't wonder that you are surprised," said she, and I could see that her fingers were trembling as she undid the fastening of her mantle. "Why, I never remember having done such a thing in my life before. The fact is, that I felt as though I were choking, and had a perfect longing for a breath of fresh air. I really think that I should have fainted if I had not gone out. I stood at the door for a few minutes, and now I am quite myself again."

'All the time that she was telling me this story she never once looked in my direction, and her voice was quite unlike her usual tones. It was evident to me that she was saying what was false. I said nothing in reply, but turned my face to the wall, sick at heart, with my mind filled with a thousand venomous doubts and suspicions. What was it that my wife was concealing from me? Where had she been during that strange expedition? I felt that I should have no peace until I knew, and yet I shrank from asking her again after once she had told me what was false. All the rest of the night I tossed and tumbled, framing theory after theory, each more unlikely than the last.

'I should have gone to the City that day, but I was too perturbed in my mind to be able to pay attention to business matters. My wife seemed to be as upset as myself, and I could see from the little questioning glances which she kept shooting at me, that she understood that I disbelieved her statement, and that she was at her wits' ends what to do. We hardly exchanged a word during breakfast, and immediately afterwards I went out for a walk, that I might think the matter over in the fresh morning air.

'I went as far as the Crystal Palace,* spent an hour in the grounds, and was back in Norbury by one o'clock. It happened that my way took me past the cottage, and I stopped for an instant to look at the windows and to see if I could catch a glimpse of the strange face which had stared out at me on the day before. As I stood there, imagine my surprise, Mr Holmes, when the door suddenly opened and my wife walked out!

'I was struck dumb with astonishment at the sight of her, but my emotions were nothing to those which showed themselves upon her face when our eyes met. She seemed for an instant to wish to shrink back inside the house again, and then, seeing how useless all concealment must be, she came forward with a very white face and frightened eyes which belied the smile upon her lips.

' "Oh, Jack!" she said, "I have just been in to see if I can be of any assistance to our new neighbours. Why do you look at me like that, Jack? You are not angry with me?"

' "So," said I, "this is where you went during the night?"

' "What do you mean?" she cried.

' "You came here. I am sure of it. Who are these people that you should visit them at such an hour?"

' "I have not been here before."

' "How can you tell me what you know is false?" I cried. "Your very voice changes as you speak. When have I ever had a secret from you? I shall enter that cottage, and I shall probe the matter to the bottom."

' "No, no, Jack, for God's sake!" she gasped, in incontrollable emotion. Then as I approached the door, she seized my sleeve and pulled me back with convulsive strength.

' "I implore you not to do this, Jack," she cried. "I swear that I will tell you everything some day, but nothing but misery can come of it if you enter that cottage." Then, as I tried to shake her off, she clung to me in a frenzy of entreaty.

' "Trust me, Jack!" she cried. "Trust me only this once. You will never have cause to regret it. You know that I would not have a secret from you if it were not for your own sake. Our whole lives are at stake on this. If you come home with me all will be well. If you force your way into that cottage, all is over between us."

'There was such earnestness, such despair in her manner that her words arrested me, and I stood irresolute before the door.

' "I will trust you on one condition, and on one condition only," said I at last. "It is that this mystery comes to an end from now. You are at liberty to preserve your secret, but you must promise me that there shall be no more nightly visits, no more doings which are kept from my knowledge. I am willing to forget those which are passed if you will promise that there shall be no more in the future."

' "I was sure that you would trust me," she cried, with a great sigh of relief. "It shall be just as you wish. Come away, oh, come away up to the house!" Still plucking at my sleeve she led me away from the cottage. As we went I glanced back, and there was that yellow, livid face watching us out of the upper window. What link could there be between that creature and my wife? Or how could the coarse, rough woman whom I had seen the day before be connected with her? It was a strange puzzle, and yet I knew that my mind could never know ease again until I had solved it.

'For two days after this I stayed at home, and my wife appeared to abide loyally by our engagement, for, as far as I know, she never stirred out of the house. On the third day, however, I had ample evidence that her solemn promise was not enough to hold her back from this secret influence which drew her away from her husband and her duty.

'I had gone into town on that day, but I returned by the 2.40 instead of the 3.36, which is my usual train. As I entered the house the maid ran into the hall with a startled face.

' "Where is your mistress?" I asked.

' "I think that she has gone out for a walk," she answered.

'My mind was instantly filled with suspicion. I rushed upstairs to make sure that she was not in the house. As I did so I happened to glance out of one of the upper windows, and saw the maid with whom I had just been speaking running across the field in the direction of the cottage. Then, of course, I saw exactly what it all meant. My wife had gone over there and had asked the servant to call her if I should return. Tingling with anger, I rushed down and strode across, determined to end the matter once and for ever. I saw my wife and the maid hurrying back together along the lane, but I did not stop to speak with them. In the cottage lay the secret which was casting a shadow over my life. I vowed that, come what might, it should be a secret no longer. I did not even knock when I reached it, but turned the handle and rushed into the passage.

'It was all still and quiet upon the ground-floor. In the kitchen a kettle was singing on the fire, and a large black cat lay coiled up in a basket, but there was no sign of the woman whom I had seen before. I ran into the other room, but it was equally deserted. Then I rushed up the stairs, but only to find two other rooms empty and deserted at the top. There was no one at all in the whole house. The furniture and pictures were of the most common and vulgar description, save in the one chamber at the window of which I had seen the strange face. That was comfortable and elegant, and all my suspicions rose into a fierce, bitter blaze when I saw that on the mantelpiece stood a full-length photograph of my wife, which had been taken at my request only three months ago.

'I stayed long enough to make certain that the house was absolutely empty. Then I left it, feeling a weight at my heart such as I had never had before. My wife came out into the hall as I entered my house, but I was too hurt and angry to speak with her, and pushing past her I made my way into my study. She followed me, however, before I could close the door.

' "I am sorry that I broke my promise, Jack," said she, "but if you knew all the circumstances I am sure you would forgive me."

' "Tell me everything, then," said I.

' "I cannot, Jack, I cannot!" she cried.

' "Until you tell me who it is that has been living in that cottage, and who it is to whom you have given that photograph, there can never be any confidence between us," said I, and breaking away from her I left the house. That was yesterday, Mr Holmes, and I have not seen her since, nor do I know anything more about this strange business. It is the first shadow that has come between us, and it has so shaken me that I do not know what I should do for the best. Suddenly this morning it occurred to me that you were the man to advise me, so I have hurried to you now, and I place myself unreservedly in your hands. If there is any point which I have not made clear, pray question me about it. But above all tell me quickly what I have to do, for this misery is more than I can bear.'

Holmes and I had listened with the utmost interest to this extraordinary statement, which had been delivered in the jerky, broken fashion of a man who is under the influence of extreme emotion. My companion sat silent now for some time, with his chin upon his hand, lost in thought.

'Tell me,' said he at last, 'could you swear that this was a man's face which you saw at the window?'

'Each time that I saw it I was some distance away from it, so that it is impossible for me to say.'

'You appear, however, to have been disagreeably impressed by it.'

'It seemed to be of an unnatural colour and to have a strange rigidity about the features. When I approached, it vanished with a jerk.'

'How long is it since your wife asked you for a hundred pounds?'

'Nearly two months.'

'Have you ever seen a photograph of her first husband?'

'No; there was a great fire at Atlanta very shortly after his death, and all her papers were destroyed.'

'And yet she had a certificate of death. You say that you saw it?'

'Yes, she got a duplicate after the fire.'

'Did you ever meet anyone who knew her in America?'

'No.'

'Did she ever talk of revisiting the place?'

'No.'

'Or get letters from it?'

'Not to my knowledge.'

'Thank you. I should like to think over the matter a little now. If the cottage is permanently deserted we may have some difficulty; if on the other hand, as I fancy is more likely, the inmates were warned of your coming, and left before you entered yesterday, then they may be back now, and we should clear it all up easily. Let me advise you, then, to return to Norbury and to examine the windows of the cottage again. If you have reason to believe that it is inhabited do not force your way in, but send a wire to my friend and me. We shall be with you within an hour of receiving it, and we shall then very soon get to the bottom of the business.'

'And if it is still empty?'

'In that case I shall come out to-morrow and talk it over with you. Good-bye, and above all things do not fret until you know that you really have a cause for it.'

'I am afraid that this is a bad business, Watson,' said my companion, as he returned after accompanying Mr Grant Munro to the door. 'What do you make of it?'

'It has an ugly sound,' I answered.

'Yes. There's blackmail in it, or I am much mistaken.'

'And who is the blackmailer?'

'Well, it must be this creature who lives in the only comfortable room in the place, and has her photograph above his fireplace. Upon my word, Watson, there is something very attractive about that livid face at the window, and I would not have missed the case for worlds.'

'You have a theory?'

'Yes, a provisional one. But I shall be surprised if it does not turn out to be correct. This woman's first husband is in that cottage.'

'Why do you think so?'

'How else can we explain her frenzied anxiety that her second one should not enter it? The facts, as I read them, are something like this: This woman was married in America. Her husband developed some hateful qualities, or, shall we say, that he contracted some loathsome disease, and became a leper or an imbecile. She fled from him at last, returned to England, changed her name, and started her life, as she thought, afresh. She had been married three years, and believed that her position was quite secure—having shown her husband the death certificate of some man, whose name she had assumed—when suddenly her whereabouts was discovered by her first husband, or, we may suppose, by some unscrupulous woman, who had attached herself to the invalid. They write to the wife and threaten to come and expose her. She asks for a hundred pounds and endeavours to buy them off. They come in spite of it, and when the husband mentions casually to the wife that there are new-comers in the cottage, she knows in some way that they are her pursuers. She waits until her husband is asleep, and then she rushes down to endeavour to persuade them to leave her in peace. Having no success, she goes again next morning, and her husband meets her, as he has told us, as she came out. She promises him then not to go there again, but two days afterwards, the hope of getting rid of those dreadful neighbours is too strong for her, and she makes another attempt, taking down with her the photograph which had probably been demanded from her. In the midst of this interview the maid rushes in to say that the master has come home, on which the wife, knowing that he would come straight down to the cottage, hurries the inmates out at the back door, into that grove of fir trees probably which was mentioned as standing near. In this way he finds the place deserted. I shall be very much surprised, however, if it is still so when he reconnoitres it this evening. What do you think of my theory?'

'It is all surmise.'

'But at least it covers all the facts. When new facts come to our knowledge which cannot be covered by it, it will be time enough to reconsider it. At present we can do nothing until we have a fresh message from our friend at Norbury.'

But we had not very long to wait. It came just as we had finished our tea. 'The cottage is still tenanted,' it said. 'Have seen the face again at the window. I'll meet the seven o'clock train, and take no steps until you arrive.'

He was waiting on the platform when we stepped out, and we could see in the light of the station lamps that he was very pale, and quivering with agitation.

'They are still there, Mr Holmes,' said he, laying his hand upon my friend's sleeve. 'I saw lights in the cottage as I came down. We shall settle it now, once and for all.'

'What is your plan, then?' asked Holmes, as we walked down the dark, tree-lined road.

'I am going to force my way in and see for myself who is in the house. I wish you both to be there as witnesses.'

'You are quite determined to do this, in spite of your wife's warning that it is better that you should not solve the mystery?'

'Yes, I am determined.'

'Well, I think that you are in the right. Any truth is better than indefinite doubt. We had better go up at once. Of course, legally we are putting ourselves hopelessly in the wrong, but I think that it is worth it.'

It was a very dark night and a thin rain began to fall as we turned from the high road into a narrow lane, deeply rutted, with hedges on either side. Mr Grant Munro pushed impatiently forward, however, and we stumbled after him as best we could.

'There are the lights of my house,' he murmured, pointing to a glimmer among the trees, 'and here is the cottage which I am going to enter.'

We turned a corner in the lane as he spoke, and there was the building close beside us. A yellow bar falling across the black foreground showed that the door was not quite closed, and one window in the upper story was brightly illuminated. As we looked we saw a dark blur* moving across the blind.

'There is that creature,' cried Grant Munro; 'you can see for yourselves that someone is there. Now follow me, and we shall soon know all.'

We approached the door, but suddenly a woman appeared out of the shadow and stood in the golden track of the lamp-light. I could not see her face in the darkness, but her arms were thrown out in an attitude of entreaty.

'For God's sake, don't, Jack!' she cried. 'I had a presentiment that you would come this evening. Think better of it, dear! Trust me again, and you will never have cause to regret it.'

'I have trusted you too long, Effie!' he cried, sternly. 'Leave go of me! I must pass you. My friends and I are going to settle this matter once and for ever.' He pushed her to one side and we followed closely after him. As he threw the door open an elderly woman ran out in front of him and tried to bar his passage, but he thrust her back, and an instant afterwards we were all upon the stairs. Grant Munro rushed into the lighted room at the top, and we entered it at his heels.

It was a cosy, well-furnished apartment, with two candles burning upon the table and two upon the mantelpiece. In the corner, stooping over a desk, there sat what appeared to be a little girl. Her face was turned away as we entered, but we could see that she was dressed in a red frock, and that she had long white gloves on. As she whisked round to us I gave a cry of surprise and horror. The face which she turned towards us was of the strangest livid tint, and the features were absolutely devoid of any expression. An instant later the mystery was explained. Holmes, with a laugh, passed his hand behind the child's ear, a mask peeled off from her countenance, and there was a little coal-black negress with all her white teeth flashing in amusement at our amazed faces. I burst out laughing out of sympathy with her merriment, but Grant Munro stood staring, with his hand clutching at his throat.

'My God!' he cried, 'what can be the meaning of this?'

'I will tell you the meaning of it,' cried the lady, sweeping into the room with a proud, set face. 'You have forced me against my own judgment to tell you, and now we must both make the best of it. My husband died at Atlanta. My child survived.'

'Your child!'

She drew a large silver locket from her bosom. 'You have never seen this open.'

'I understood that it did not open.'

She touched a spring, and the front hinged back. There was a portrait within of a man, strikingly handsome and intelligent, but bearing unmistakable signs upon his features of his African descent.

'That is John Hebron, of Atlanta,' said the lady, 'and a nobler man never walked the earth. I cut myself off from my race in order to wed him; but never once while he lived did I for one instant regret it. It was our misfortune that our only child took after his people rather than mine. It is often so in such matches, and little Lucy is darker far than ever her father was.* But, dark or fair, she is my own dear little girlie, and her mother's pet.' The little creature ran across at the words and nestled up against the lady's dress.

'When I left her in America,' she continued, 'it was only because her health was weak, and the change might have done her harm. She was given to the care of a faithful Scotchwoman who had once been our servant. Never for an instant did I dream of disowning her as my child. But when chance threw you in my way, Jack, and I learned to love you, I feared to tell you about my child. God forgive me, I feared that I should lose you, and I had not the courage to tell you. I had to choose between you, and in my weakness I turned away from my own little girl. For three years I have kept her existence a secret from you, but I heard from the nurse, and I knew that all was well with her. At last, however, there came an overwhelming desire to see the child once more. I struggled against it, but in vain. Though I knew the danger I determined to have the child over, if it were but for a few weeks. I sent a hundred pounds to the nurse, and I gave her instructions about this cottage, so that she might come as a neighbour without my appearing to be in any way connected with her. I pushed my precautions so far as to order her to keep the child in the house during the daytime, and to cover up her little face and hands, so that even those who might see her at the window should not gossip about there being a black child in the neighbourhood. If I had been less cautious I might have been more wise, but I was half crazy with fear lest you should learn the truth.

'It was you who told me first that the cottage was occupied. I should have waited for the morning, but I could not sleep for excitement, and so at last I slipped out, knowing how difficult it is to awaken you. But you saw me

71

go, and that was the beginning of my troubles. Next day you had my secret at your mercy, but you nobly refrained from pursuing your advantage. Three days later, however, the nurse and child only just escaped from the back door as you rushed in at the front one. And now to-night you at last know all, and I ask you what is to become of us, my child and me?' She clasped her hands and waited for an answer.

It was a long two* minutes before Grant Munro broke the silence, and when his answer came it was one of which I love to think. He lifted the little child, kissed her, and then, still carrying her, he held his other hand out to his wife, and turned towards the door.

'We can talk it over more comfortably at home,' said he. 'I am not a very good man, Effie, but I think that I am a better one than you have given me credit for being.'

Holmes and I followed them down to the lane, and my friend plucked at my sleeve as we came out. 'I think,' said he, 'that we shall be of more use in London than in Norbury.'

Not another word did he say of the case until late that night when he was turning away, with his lighted candle, for his bedroom.

'Watson,' said he, 'if it should ever strike you that I am getting a little over-confident in my powers, or giving less pains to a case than it deserves, kindly whisper "Norbury" in my ear, and I shall be infinitely obliged to you.'

The Stockbroker's Clerk

SHORTLY after my marriage I had bought a connection* in the Paddington district. Old Mr Farquhar, from whom I purchased it, had at one time an excellent general practice, but his age, and an affliction of the nature of St Vitus's Dance* from which he suffered, had very much thinned it. The public, not unnaturally, goes upon the principle that he who would heal others must himself be whole, and looks askance at the curative powers of the man whose own case is beyond the reach of his drugs. Thus, as my predecessor weakened, his practice declined, until when I purchased it from him it had sunk from twelve hundred to little more than three hundred a year.* I had confidence, however, in my own youth and energy, and was convinced that in a very few years the concern would be as flourishing as ever.

For three months after taking over the practice I was kept very closely at work, and saw little of my friend Sherlock Holmes, for I was too busy to visit Baker Street, and he seldom went anywhere himself save upon professional business. I was surprised, therefore, when one morning in June, as I sat reading the *British Medical Journal** after break-fast, I heard a ring at the bell followed by the high, somewhat strident, tones of my old companion's voice.

'Ah, my dear Watson,' said he, striding into the room, 'I am very delighted to see you. I trust that Mrs Watson has entirely recovered from all the little excitements connected with our adventure of the "Sign of Four"*?'

'Thank you, we are both very well,' said I, shaking him warmly by the hand.

'And I hope also,' he continued, sitting down in the rocking-chair, 'that the cares of medical practice have not entirely obliterated the interest which you used to take in our little deductive problems.'

'On the contrary,' I answered; 'it was only last night that I was looking over my old notes and classifying some of our past results.'

'I trust that you don't consider your collection closed?'

'Not at all. I should wish nothing better than to have some more of such experiences.'

'To-day, for example?'

'Yes; to-day, if you like.'

'And as far off as Birmingham?'

'Certainly, if you wish it.'

'And the practice?'

'I do my neighbour's when he goes. He is always ready to work off the debt.'

'Ha! Nothing could be better!' said Holmes, leaning back in his chair and looking keenly at me from under his half-closed lids. 'I perceive that you have been unwell lately. Summer colds are always a little trying.'

'I was confined to the house by a severe chill for three days last week. I thought, however, that I had cast off every trace of it.'

'So you have. You look remarkably robust.'

'How, then, did you know of it?'

'My dear fellow, you know my methods.'

'You deduced it, then?'

'Certainly.'

'And from what?'

'From your slippers.'

I glanced down at the new patent leathers which I was wearing. 'How on earth—?' I began, but Holmes answered my question before it was asked.

'Your slippers are new,' he said. 'You could not have had them more than a few weeks. The soles which you are at this moment presenting to me are slightly scorched. For a moment I thought they might have got wet and been burned in the drying. But near the instep there is a small circular wafer of paper with the shopman's hieroglyphics upon it. Damp would of course have removed this. You had then been sitting with your feet outstretched to the fire, which a man would hardly do even in so wet a June as this if he were in his full health.'

Like all Holmes's reasoning the thing seemed simplicity itself when it was once explained. He read the thought upon my features, and his smile had a tinge of bitterness.

'I am afraid that I rather give myself away when I explain,' said he. 'Results without causes are much more impressive. You are ready to come to Birmingham, then?'

'Certainly. What is the case?'

'You shall hear it all in the train. My client is outside in a four-wheeler.* Can you come at once?'

'In an instant.' I scribbled a note to my neighbour, rushed upstairs to explain the matter to my wife, and joined Holmes upon the doorstep.

'Your neighbour is a doctor?' said he, nodding at the brass plate.

'Yes. He bought a practice as I did.'

'An old-established one?'

'Just the same as mine. Both have been ever since the houses were built.'

'Ah, then you got hold of the best of the two.'

'I think I did. But how do you know?'

'By the steps, my boy. Yours are worn three inches deeper than his. But this gentleman in the cab is my client, Mr Hall Pycroft. Allow me to introduce you to him. Whip your horse up, cabby, for we have only just time to catch our train.'

The man whom I found myself facing was a well-built, fresh-complexioned young fellow with a frank, honest face and a slight, crisp, yellow moustache. He wore a very shiny top-hat and a neat suit of sober black, which made him look what he was—a smart young City man, of the class who have been labelled Cockneys,* but who give us our crack Volunteer regiments,* and who turn out more fine athletes and sportsmen than any body of men in these islands. His round, ruddy face was naturally full of cheeriness, but the corners of his mouth seemed to me to be pulled down in a half-comical distress. It was not, however, until we were all in a first-class carriage and well started upon our journey to

Birmingham, that I was able to learn what the trouble was which had driven him to Sherlock Holmes.

'We have a clear run here of seventy minutes,' Holmes remarked. 'I want you, Mr Hall Pycroft, to tell my friend your very interesting experience exactly as you have told it to me, or with more detail if possible. It will be of use to me to hear the succession of events again. It is a case, Watson, which may prove to have something in it, or may prove to have nothing, but which at least presents those unusual and *outré** features which are as dear to you as they are to me. Now, Mr Pycroft, I shall not interrupt you again.'

Our young companion looked at me with a twinkle in his eye.

'The worst of the story is,' said he, 'that I show myself up as such a confounded fool. Of course, it may work out all right, and I don't see that I could have done otherwise; but if I have lost my crib* and get nothing in exchange, I shall feel what a soft Johnny I have been. I'm not very good at telling a story, Dr Watson, but it is like this with me.

'I used to have a billet at Coxon and Woodhouse, of Drapers' Gardens, but they were let in* early in the spring through the Venezuelan loan,* as no doubt you remember, and came a nasty cropper.* I had been with them five years, and old Coxon gave me a ripping good testimonial when the smash came; but, of course, we clerks were all turned adrift, the twenty-seven of us. I tried here and tried there, but there were lots of other chaps on the same lay* as myself, and it was a perfect frost for a long time. I had been taking three pounds a week at Coxon's, and I had saved about seventy of them, but I soon worked my way through that and out at the other end. I was fairly at the end of my tether at last, and could hardly find the stamps to answer the advertisements or the envelopes to stick them to. I had worn out my boots padding up office stairs, and I seemed just as far from getting a billet as ever.

'At last I saw a vacancy at Mawson and Williams', the great stockbroking firm in Lombard Street.* I dare say EC* is not much in your line, but I can tell you that this is about the richest house in London. The advertisement was to be answered by letter only. I sent in my testimonial and application, but without the least hope of getting it. Back came an answer by return saying that if I would appear next Monday I might take over my new duties at once, provided that my appearance was satisfactory. No one knows how these things are worked. Some people say the manager just plunges his hand into the heap and takes the first that comes. Anyhow, it was my innings that time, and I don't ever wish to feel better pleased. The screw* was a pound a week rise, and the duties just about the same as at Coxon's.

'And now I come to the queer part of the business. I was in diggings out Hampstead way—17, Potter's Terrace, was the address. Well, I was sitting doing a smoke that very evening after I had been promised the appointment, when up came my landlady with a card which had "Arthur Pinner, financial agent," printed upon it. I had never heard the name before, and could not imagine what he wanted with me, but of course I asked her to show him up. In he walked—a middle-sized, dark-haired, dark-eyed, black-bearded man, with a touch of the sheeny* about his nose. He had a brisk kind of way with him and spoke sharply, like a man that knew the value of time.

' "Mr Hall Pycroft, I believe?" said he.

' "Yes, sir," I answered, and pushed a chair towards him.

' "Lately engaged at Coxon and Woodhouse's?"

' "Yes, sir."

' "And now on the staff of Mawson's?"

' "Quite so."

' "Well," said he. "The fact is that I have heard some really extraordinary stories about your financial ability. You remember Parker who used to be Coxon's manager? He can never say enough about it."

'Of course I was pleased to hear this. I had always been pretty smart in the office, but I had never dreamed that I was talked about in the City in this fashion.

' "You have a good memory?" said he.

' "Pretty fair," I answered, modestly.

' "Have you kept in touch with the market while you have been out of work?" he asked.

' "Yes; I read the Stock Exchange List* every morning."

' "Now, that shows real application!" he cried. "That is the way to prosper! You won't mind my testing you, will you? Let me see! How are Ayrshires?"*

' "One hundred and five, to one hundred and five and a quarter,"* I answered.

' "And New Zealand Consolidated?"

' "A hundred and four."

' "And British Broken Hills?"*

' "Seven to seven and six."

' "Wonderful!" he cried, with his hands up. "This quite fits in with all that I had heard. My boy, my boy, you are very much too good to be a clerk at Mawson's!"

'This outburst rather astonished me, as you can think. "Well," said I, "other people don't think quite so much of me as you seem to do, Mr Pinner. I had a hard enough fight to get this berth, and I am very glad to have it."

' "Pooh, man, you should soar above it. You are not in your true sphere. Now, I'll tell you how it stands with me. What I have to offer is little enough when measured by your ability, but when compared with Mawson's it is light to dark. Let me see! When do you go to Mawson's?"

' "On Monday."

' "Ha! ha! I think I would risk a little sporting flutter that you don't go there at all."

' "Not go to Mawson's?"

· ' "No, sir. By that day you will be the business manager of the Franco-Midland Hardware Company, Limited, with one hundred and thirty-four branches in the towns and villages of France, not counting one in Brussels and one in San Remo."

'This took my breath away. "I never heard of it," said I.

' "Very likely not. It has been kept very quiet, for the capital was all privately subscribed, and it is too good a thing to let the public into. My brother, Harry Pinner, is promoter, and joins the board after allotment as managing director. He knew that I was in the swim down here, and he asked me to pick up a good man cheap—a young pushing man, with plenty of snap about him. Parker spoke of you, and that brought me here tonight. We can only offer you a beggarly five hundred to start with—"

' "Five hundred a year!" I shouted.

' "Only that at the beginning, but you are to have an over-riding commission of 1 per cent on all business done by your agents, and you may take my word for it that this will come to more than your salary."

' "But I know nothing about hardware."

' "Tut, my boy, you know about figures."

'My head buzzed, and I could hardly sit still in the chair. But suddenly a little chill of doubt came over me.

' "I must be frank with you," said I. "Mawson only gives me two hundred, but Mawson is safe. Now, really, I know so little about your company that—"

' "Ah, smart, smart!" he cried, in a kind of ecstasy of delight. "You are the very man for us! You are not to be talked over, and quite right too. Now here's a note for a hundred pounds; and if you think that we can do business you may just slip it into your pocket as an advance upon your salary."

' "That is very handsome," said I. "When shall I take over my new duties?"

' "Be in Birmingham tomorrow at one," said he. "I have a note in my pocket here which you will take to my brother. You will find him at 126B, Corporation Street, where the temporary offices of the company are situated. Of course he must confirm your engagement, but between ourselves it will be all right."

' "Really, I hardly know how to express my gratitude, Mr Pinner," said I.

' "Not at all, my boy. You have only got your deserts. There are one or two small things—mere formalities—which I must arrange with you. You have a bit of paper beside you there. Kindly write upon it, 'I am perfectly willing to act as business manager to the Franco-Midland Hardware Company, Limited, at a minimum salary of £500.' "

'I did as he asked, and he put the paper in his pocket.

' "There is one other detail," said he. "What do you intend to do about Mawson's?"

'I had forgotten all about Mawson's in my joy.

' "I'll write and resign," said I.

' "Precisely what I don't want you to do. I had a row over you with Mawson's manager. I had gone up to ask him about you, and he was very offensive—accused me of coaxing you away from the service of the firm, and that sort of thing. At last I fairly lost my temper. 'If you want good men you should pay them a good price,' said I. 'He would rather have our small price than your big one,' said he. 'I'll lay you a fiver,'* said I, 'that when he has my offer you will never so much as hear from him again.' 'Done!' said he. 'We picked him out of the gutter, and he won't leave us so easily.' Those were his very words."

' "The impudent scoundrel!" I cried. "I've never so much as seen him in my life. Why should I consider him in any way? I shall certainly not write if you would rather that I didn't."

' "Good! That's a promise!" said he, rising from his chair. "Well, I am delighted to have got so good a man for my brother. Here is your advance of a hundred pounds, and here is the letter. Make a note of the address, 126B, Corporation Street, and remember that one o'clock tomorrow is your appointment. Good-night, and may you have all the fortune that you deserve."

'That's just about all that passed between us as near as I can remember it. You can imagine, Dr Watson, how pleased I was at such an extraordinary bit of good fortune. I sat up half the night hugging myself over it, and next day I was off to Birmingham in a train that would take me in plenty of time for my appointment. I took my things to an hotel in New Street,* and then I made my way to the address which had been given me.

'It was a quarter of an hour before my time, but I thought that would make no difference. 126B was a passage between two large shops which led to a winding stone stair, from which there were many flats, let as offices to companies or professional men. The names of the occupants were painted up at the bottom on the wall, but there was no such name as the Franco-Midland Hardware Company, Limited. I stood for a few minutes with my heart in my boots, wondering whether the whole thing was an elaborate hoax or not, when up came a man and addressed me. He was very like the chap that I had seen the night before, the same figure and voice, but he was clean shaven and his hair was lighter.

' "Are you Mr Hall Pycroft?" he asked.

' "Yes," said I.

' "Ah! I was expecting you, but you are a trifle before your time. I had a note from my brother this morning, in which he sang your praises very loudly."

' "I was just looking for the offices when you came."

' "We have not got our name up yet, for we only secured these temporary premises last week. Come up with me and we will talk the matter over."

'I followed him to the top of a very lofty stair, and there right under the slates were a couple of empty and dusty little rooms, uncarpeted and uncurtained, into which he led me. I had thought of a great office with shining tables and rows of clerks such as I was used to, and I dare say I stared rather straight at the two deal* chairs and one little table, which, with a ledger and a waste-paper basket, made up the whole furniture.

' "Don't be disheartened, Mr Pycroft," said my new acquaintance, seeing the length of my face. "Rome was not built in a day, and we have lots of money at our backs, though we don't cut much dash yet in offices. Pray sit down and let me have your letter."

'I gave it to him, and he read it over very carefully.

' "You seem to have made a vast impression upon my brother, Arthur," said he, "and I know that he is a pretty shrewd judge. He swears by London, you know, and I by Birmingham, but this time I shall follow his advice. Pray consider yourself definitely engaged."

' "What are my duties?" I asked.

' "You will eventually manage the great depôt in Paris, which will pour a flood of English crockery into the shops of one hundred and thirty-four agents in France. The purchase will be completed in a week, and meanwhile you will remain in Birmingham and make yourself useful."

' "How?"

'For answer he took a big red book out of a drawer. "This is a directory of Paris," said he, "with the trades after the names of the people. I want you to take it home with you, and to mark off all the hardware sellers with their addresses. It would be of the greatest use to me to have them."

' "Surely there are classified lists?" I suggested.

' "Not reliable ones. Their system is different to ours. Stick at it and let me have the lists by Monday, at twelve. Good day, Mr Pycroft; if you continue to show zeal and intelligence, you will find the company a good master."

'I went back to the hotel with the big book under my arm, and with very conflicting feelings in my breast. On the one hand I was definitely engaged, and had a hundred pounds in my pocket. On the other, the look of the offices, the absence of name on the wall, and other of the points which would strike a business man had left a bad impression as to the position of my employers. However, come what might, I had my money, so I settled down to my task. All Sunday I was kept hard at work, and yet by Monday I had only got as far as H. I went round to my employer, found him in the same dismantled kind of room, and was told to keep at it until Wednesday, and then come again. On Wednesday it was still unfinished, so I hammered away until Friday—that is, yesterday. Then I brought it round to Mr Harry Pinner.

' "Thank you very much," said he. "I fear that I underrated the difficulty of the task. This list will be of very material assistance to me."

' "It took some time," said I.

' "And now," said he, "I want you to make a list of the furniture shops, for they all sell crockery."

' "Very good."

' "And you can come up to-morrow evening at seven, and let me know how you are getting on. Don't overwork yourself. A couple of hours at Day's Music-Hall* in the evening would do you no harm after your labours." He laughed as he spoke, and I saw with a thrill that his second tooth upon the left-hand ide had been very badly stuffed* with gold.'

Sherlock Holmes rubbed his hands with delight, and I stared in astonishment at our client.

'You may well look surprised, Dr Watson, but it is this way,' said he. 'When I was speaking to the other chap in London at the time that he laughed at my not going to Mawson's, I happened to notice that his tooth was stuffed in this very identical fashion. The glint of the gold in each case caught my eye, you see. When I put that with the voice and figure being the same, and only those things altered which might be changed by a razor or a wig, I could not doubt that it was the same man. Of course, you expect two brothers to be alike, but not that they should have the same tooth stuffed in the same way. He bowed me out and I found myself in the street, hardly knowing whether I was on my head or my heels. Back I went to my hotel, put my head in a basin of cold water, and tried to think it out. Why had he sent me from London to Birmingham; why had he got there before me; and why had he written a letter from himself to himself? It was altogether too much for me, and I could make no sense of it. And then suddenly it struck me that what was dark to me might be very light to Mr Sherlock Holmes. I had just time to get up to town by the night train, to see him this morning, and to bring you both back with me to Birmingham.'

There was a pause after the stockbroker's clerk had concluded his surprising experience. Then Sherlock Holmes cocked his eye at me, leaning back on the cushions with a pleased and yet critical face, like a connoisseur who had just taken his first sip of a comet vintage.*

'Rather fine, Watson, is it not?' said he. 'There are points in it which please me. I think you will agree with me that an interview with Mr Arthur Harry Pinner* in the temporary offices of the Franco-Midland Hardware Company, Limited, would be a rather interesting experience for both of us.'

'But how can we do it?' I asked.

'Oh, easily enough,' said Hall Pycroft, cheerily. 'You are two friends of mine who are in want of a billet, and what could be more natural than that I should bring you both round to the managing director?'

'Quite so! Of course!' said Holmes. 'I should like to have a look at the gentleman and see if I can make anything of his little game. What qualities have you, my friend, which would make your services so valuable? or is it possible that—' He began biting his nails and staring blankly out of the window, and we hardly drew another word from him until we were in New Street.

At seven o'clock that evening we were walking, the three of us, down Corporation Street to the company's offices.

'It is of no use our being at all before our time,' said our client. 'He only comes there to see me apparently, for the place is deserted up to the very hour he names.'

'That is suggestive,' remarked Holmes.

'By Jove, I told you so!' cried the clerk. 'That's he walking ahead of us there.'

He pointed to a smallish, blond, well-dressed man, who was bustling along the other side of the road. As we watched him he looked across at a boy who was bawling out the latest edition of the evening paper, and, running over among the cabs and 'buses, he bought one from him. Then clutching it in his hand he vanished through a doorway.

'There he goes!' cried Hall Pycroft. 'Those are the company's offices into which he has gone. Come with me and I'll fix it up as easily as possible.'

Following his lead we ascended five stories, until we found ourselves outside a half-opened door, at which our client tapped. A voice within bade us 'come in,' and we entered a bare, unfurnished room, such as Hall Pycroft had described. At the single table sat the man whom we had seen in the street, with his evening paper spread out in front of him, and as he looked up at us it seemed to me that I had never looked upon a face which bore such marks of grief, and of something beyond grief—of a horror such as comes to few men in a lifetime. His brow glistened with perspiration, his cheeks were of the dull dead white of a fish's belly, and his eyes were wild and staring. He looked at his clerk as though he failed to recognize him, and I could see, by the astonishment depicted upon our conductor's face, that this was by no means the usual appearance of his employer.

'You look ill, Mr Pinner,' he exclaimed.

'Yes, I am not very well,' answered the other, making obvious efforts to pull himself together, and licking his dry lips before he spoke. 'Who are these gentlemen whom you have brought with you?'

'One is Mr Harris, of Bermondsey, and the other is Mr Price, of this town,' said our clerk glibly. 'They are friends of mine, and gentlemen of experience, but they have been out of a place for some little time, and they hoped that perhaps you might find an opening for them in the company's employment.'

'Very possibly! Very possibly!' cried Mr Pinner, with a ghastly smile. 'Yes, I have no doubt that we shall be able to do something for you. What is your particular line, Mr Harris?'

'I am an accountant,' said Holmes.

'Ah, yes, we shall want something of the sort. And you, Mr Price?'

'A clerk,' said I.

'I have every hope that the company may accommodate you. I will let you know about it as soon as we come to any conclusion. And now I beg that you will go. For God's sake, leave me to myself!'

These last words were shot out of him, as though the constraint which he was evidently setting upon himself had suddenly and utterly burst asunder. Holmes and I glanced at each other, and Hall Pycroft took a step towards the table.

'You forget, Mr Pinner, that I am here by appointment to receive some directions from you,' said he.

'Certainly, Mr Pycroft, certainly,' the other answered in a calmer tone. 'You may wait here a moment, and there is no reason why your friends should not wait with you. I will be entirely at your service in three minutes, if I might trespass upon your patience so far.' He rose with a very courteous air, and bowing to us he passed out through a door at the further end of the room, which he closed behind him.

'What now?' whispered Holmes. 'Is he giving us the slip?'

'Impossible,' answered Pycroft.

'Why so?'

'That door leads into an inner room.'

'There is no exit?'

'None.'

'Is it furnished?'

'It was empty yesterday.'

'Then what on earth can he be doing? There is something which I don't understand in this matter. If ever a man was three parts mad with terror, that man's name is Pinner. What can have put the shivers on him?'

'He suspects that we are detectives,' I suggested.

'That's it,' said Pycroft.

Holmes shook his head. 'He did not turn pale. He *was* pale when we entered the room,' said he. 'It is just possible that—'

His words were interrupted by a sharp rat-tat from the direction of the inner door.

'What the deuce is he knocking at his own door for?' cried the clerk.

Again and much louder came the rat-tat-tat. We all gazed expectantly at the closed door. Glancing at Holmes I saw his face turn rigid, and he leaned forward in intense excitement. Then suddenly came a low gurgling, gargling sound and a brisk drumming upon woodwork. Holmes sprang frantically across the room and pushed at the door. It was fastened on the inner side. Following his example, we threw ourselves upon it with all our weight. One hinge snapped, then the other, and down came the door with a crash. Rushing over it we found ourselves in the inner room.

It was empty.

But it was only for a moment that we were at fault. At one corner, the corner nearest the room which we had left, there was a second door. Holmes sprang to it and pulled it open. A coat and waistcoat were lying on the floor, and from a hook behind the door, with his own braces round his neck, was hanging the managing director of the Franco-Midland Hardware Company. His knees were drawn up, his head hung at a dreadful angle to his body, and the clatter of his heels against the door made the noise which had broken in upon our conversation. In an instant I had caught him round the waist and held him up, while Holmes and Pycroft untied the elastic bands which had disappeared between the livid creases of skin. Then we carried him into the other room, where he lay with a slate-coloured face,* puffing his purple lips in and out with every breath—a dreadful wreck of all that he had been but five minutes before.

'What do you think of him, Watson?' asked Holmes.

I stooped over him and examined him. His pulse was feeble and intermittent, but his breathing grew longer, and there was a little shivering of his eyelids which showed a thin white slit of ball beneath.

'It has been touch and go with him,' said I, 'but he'll live now. Just open that window and hand me the water carafe.' I undid his collar, poured the cold water over his face, and raised and sank his arms until he drew a long natural breath.

'It's only a question of time now,' said I, as I turned away from him.

Holmes stood by the table with his hands deep in his trousers pockets and his chin upon his breast.

'I suppose we ought to call the police in now,' said he; 'and yet I confess that I like to give them a complete case when they come.'

87

'It's a blessed mystery to me,' cried Pycroft, scratching his head. 'Whatever they wanted to bring me all the way up here for, and then—'

'Pooh! All that is clear enough,' said Holmes impatiently. 'It is this last sudden move.'

'You understand the rest, then?'

'I think that is fairly obvious. What do you say, Watson?' I shrugged my shoulders.

'I must confess that I am out of my depths,' said I.

'Oh, surely, if you consider the events at first they can only point to one conclusion.'

'What do you make of them?'

'Well, the whole thing hinges upon two points. The first is the making of Pycroft write a declaration by which he entered the service of this preposterous company. Do you not see how very suggestive that is?'

'I am afraid I miss the point.'

'Well, why did they want him to do it? Not as a business matter, for these arrangements are usually verbal, and there was no earthly business reason why this should be an exception. Don't you see, my young friend, that they were very anxious to obtain a specimen of your handwriting, and had no other way of doing it?'

'And why?'

'Quite so. Why? When we answer that, we have made some progress with our little problem. Why? There can be only one adequate reason. Someone wanted to learn to imitate your writing, and had to procure a specimen of it first. And now if we pass on to the second point, we find that each throws light upon the other. That point is the request made by Pinner that you should not resign your place, but should leave the manager of this important business in the full expectation that a Mr Hall Pycroft, whom he had never seen, was about to enter the office upon the Monday morning.'

'My God!' cried our client, 'what a blind beetle I have been!'

'Now you see the point about the handwriting. Suppose that someone turned up in your place who wrote a completely different hand from that in which you had applied for the vacancy, of course the game would have been up. But in the interval the rogue learnt to imitate you, and his position was therefore secure, as I presume that nobody in the office had ever set eyes upon you?'

'Not a soul,' groaned Hall Pycroft.

'Very good. Of course, it was of the utmost importance to prevent you from thinking better of it, and also to keep you from coming into contact with anyone who might tell you that your double was at work in Mawson's office. Therefore, they gave you a handsome advance on your salary, and ran you off to the Midlands, where they gave you enough work to do to prevent your going to London, where you might have burst their little game up. That is all plain enough.'

'But why should this man pretend to be his own brother?'

'Well, that is pretty clear also. There are evidently only two of them in it. The other is personating you at the office. This one acted as your engager, and then found that he could not find you an employer without admitting a third person into his plot. That he was most unwilling to do. He changed his appearance as far as he could, and trusted that the likeness, which you could not fail to observe, would be put down to a family resemblance. But for the happy chance of the gold stuffing your suspicions would probably have never been aroused.'

Hall Pycroft shook his clenched hands in the air. 'Good Lord!' he cried. 'While I have been fooled in this way, what has this other Hall Pycroft been doing at Mawson's? What should we do, Mr Holmes? Tell me what to do!'

'We must wire to Mawson's.'

'They shut at twelve on Saturdays.'

'Never mind; there may be some door-keeper or attendant—'

'Ah, yes; they keep a permanent guard there on account of the value of the securities that they hold. I remember hearing it talked of in the City.'

'Very good, we shall wire to him, and see if all is well, and if a clerk of your name is working there. That is clear enough, but what is not so clear is why at sight of us one of the rogues should instantly walk out of the room and hang himself.'

'The paper!' croaked a voice behind us. The man was sitting up, blanched and ghastly, with returning reason in his eyes, and hands which rubbed nervously at the broad red band which still encircled his throat.

'The paper! Of course!' yelled Holmes, in a paroxysm of excitement. 'Idiot that I was! I thought so much of our visit that the paper never entered my head for an instant. To be sure, the secret must lie there.' He flattened it out on the table, and a cry of triumph burst from his lips.

'Look at this, Watson!' he cried. 'It is a London paper, an early edition of the *Evening Standard*.* Here is what we want. Look at the headlines—"Crime in the City. Murder at Mawson and Williams'. Gigantic Attempted Robbery; Capture of the Criminal." Here, Watson, we are all equally anxious to hear it, so kindly read it aloud to us.'

It appeared from its position in the paper to have been the one event of importance in town, and the account of it ran in this way:

'A desperate attempt at robbery, culminating in the death of one man and the capture of the criminal, occurred this afternoon in the City. For some time back Mawson and Williams, the famous financial house, have been the guardians of securities which amount in the aggregate to a sum of considerably over a million sterling. So conscious was the manager of the responsibility which devolved upon him in consequence of the great interests at stake, that safes of the very latest construction have been employed, and an armed watchman has been left day and night in the building. It appears that last week a new clerk, named Hall Pycroft, was engaged by the firm. This person appears to have been none other than Beddington, the famous forger and cracksman,* who, with his brother, has only recently emerged from a five years' spell of penal servitude. By some means, which are not yet clear, he succeeded in winning, under a false name, this official position in the office, which he utilized in order to obtain mouldings of various locks, and a thorough knowledge of the position of the strong room and the safes.

'It is customary at Mawson's for the clerks to leave at midday on Saturday. Sergeant Tuson, of the City Police,* was somewhat surprised therefore to see a gentleman with a carpet bag come down the steps at twenty minutes past one. His suspicions being aroused, the sergeant followed the man, and with the aid of Constable Pollock succeeded, after a most desperate resistance, in arresting him. It was at once clear that a daring and gigantic robbery had been committed. Nearly a hundred thousand pounds' worth of American railway bonds, with a large amount of scrip in other mines and companies, were discovered in the bag. On examining the premises the body of the unfortunate watchman was found doubled up and thrust into the largest of the safes, where it would not have been discovered until Monday morning had it not been for the prompt action of Sergeant Tuson. The man's skull had been shattered by a blow from a poker, delivered from behind. There could be no doubt that Beddington had obtained entrance by pretending that he had left something behind him, and having murdered the watchman, rapidly rifled the large safe, and then made off with his booty. His brother, who usually works with him, has not appeared in this job, so far as can at present be ascertained, although the police are making energetic inquiries as to his whereabouts.'

'Well, we may save the police some little trouble in that direction,' said Holmes, glancing at the haggard figure huddled up by the window. 'Human nature is a strange mixture, Watson. You see that even a villain and a murderer can inspire such affection* that his brother turns to suicide when he learns that his neck is forfeited. However, we have no choice as to our action. The doctor and I will remain on guard, Mr Pycroft, if you will have the kindness to step out for the police.'

The 'Gloria Scott'

'I HAVE some papers here,' said my friend, Sherlock Holmes, as we sat one winter's night on either side of the fire, 'which I really think, Watson, it would be worth your while to glance over. These are the documents in the extraordinary case of the *Gloria Scott*, and this is the message which struck Justice of the Peace Trevor dead with horror when he read it.'

He had picked from a drawer a little tarnished cylinder, and, undoing the tape, he handed me a short note scrawled upon a half sheet of slate-grey paper.

'The supply of game for London is going steadily up,' it ran. 'Head-keeper Hudson,* we believe, has been now told to receive all orders for fly paper, and for preservation of your hen pheasant's life.'

As I glanced up from reading this enigmatical message I saw Holmes chuckling at the expression upon my face.

'You look a little bewildered,' said he.

'I cannot see how such a message as this could inspire horror. It seems to me to be rather grotesque than otherwise.'

'Very likely. Yet the fact remains that the reader, who was a fine, robust old man, was knocked clean down by it, as if it had been the butt-end of a pistol.'

'You arouse my curiosity,' said I. 'But why did you say just now that there were very particular reasons why I should study this case?'

'Because it was the first in which I was ever engaged.'

I had often endeavoured to elicit from my companion what had first turned his mind in the direction of criminal research, but I had never caught him before in a communicative humour. Now he sat forward in his arm-chair, and spread out the documents upon his knees. Then he lit his pipe and sat for some time smoking and turning them over.

'You never heard me talk of Victor Trevor?' he asked. 'He was the only friend I made during the two years that I was at college.* I was never a very sociable fellow, Watson, always rather fond of moping in my rooms and working out my own little methods of thought, so that I never mixed much with the men of my year. Bar fencing and boxing I had few athletic tastes, and then my line of study was quite distinct from that of the other fellows, so that we had no points of contact at all. Trevor was the only man I knew, and that only through the accident of his bull-terrier freezing* on to my ankle one morning as I went down to chapel.

'It was a prosaic way of forming a friendship, but it was effective. I was laid by the heels for ten days, and Trevor used to come in to inquire after me. At first it was only a minute's chat, but soon his visits lengthened, and before the end of the term we were close friends. He was a hearty, full-blooded fellow, full of spirit and energy, the very opposite to me in most respects; but we found we had some subjects in common, and it was a bond of union when I learned that he was as friendless as I. Finally, he invited me down to his father's place at Donnithorpe, in Norfolk, and I accepted his hospitality for a month of the long vacation.

'Old Trevor was evidently a man of some wealth and consideration, a J.P.* and a landed proprietor. Donnithorpe is a little hamlet just to the north of Langmere, in the country of the Broads.* The house was an old-fashioned, wide-spread, oak-beamed, brick building, with a fine lime-lined avenue leading up to it. There was excellent wild duck shooting in the fens, remarkably good fishing, a small but select library, taken over, as I understood, from a former occupant, and a tolerable cook, so that it would be a fastidious man who could not put in a pleasant month there.

'Trevor senior was a widower, and my friend was his only son. There had been a daughter, I heard, but she had died of diphtheria while on a visit to Birmingham. The father interested me extremely. He was a man of little culture, but with a considerable amount of rude strength both physically and mentally. He knew hardly any books, but he had travelled far, had seen much of the world, and had remembered all that he had learned. In person he was a thick-set, burly man with a shock of grizzled hair, a brown, weather-beaten face, and blue eyes which were keen to the verge of fierceness. Yet he had a reputation for kindness and charity in the countryside, and was noted for the leniency of his sentences from the bench.

'One evening, shortly after my arrival, we were sitting over a glass of port after dinner, when young Trevor began to talk about those habits of observation and inference which I had already formed into a system, although I had not yet appreciated the part which they were to play in my life. The old man evidently thought that his son was exaggerating in his description of one or two trivial feats which I had performed.

' "Come now, Mr Holmes," said he, laughing good-humouredly, "I'm an excellent subject, if you can deduce anything from me."

' "I fear there is not very much," I answered. "I might suggest that you have gone about in fear of some personal attack within the last twelve months."

'The laugh faded from his lips, and he stared at me in great surprise.

' "Well, that's true enough," said he. "You know, Victor," turning to his son, "when we broke up that poaching gang, they swore to knife us; and Sir Edward Hoby has actually been attacked. I've always been on my guard since then, though I have no idea how you know it."

' "You have a very handsome stick," I answered. "By the inscription,* I observed that you had not had it more than a year. But you have taken some pains to bore the head of it and pour melted lead into the hole, so as to make it a formidable weapon. I argued that you would not take such precautions unless you had some danger to fear."

' "Anything else?" he asked, smiling.

' "You have boxed a good deal in your youth."

' "Right again. How did you know it? Is my nose knocked a little out of the straight?"

' "No," said I, "It is your ears. They have the peculiar flattening and thickening which marks the boxing man."

' "Anything else?"

' "You have done a great deal of digging, by your callosities."*

' "Made all my money at the gold-fields."

' "You have been in New Zealand."

' "Right again."

' "You have visited Japan."

' "Quite true."

' "And you have been most intimately associated with someone whose initials were J.A., and whom you afterwards were eager to entirely forget."

'Mr Trevor stood slowly up, fixed his large blue eyes on me with a strange, wild stare, and then pitched forward on his face among the nutshells which strewed the cloth, in a dead faint.

'You can imagine, Watson, how shocked both his son and I were. His attack did not last long, however, for when we undid his collar and sprinkled the water from one of the finger glasses* over his face, he gave a gasp or two and sat up.

' "Ah, boys!" said he, forcing a smile. "I hope I haven't frightened you. Strong as I look, there is a weak place in my heart, and it does not take much to knock me over. I don't know how you manage this, Mr Holmes, but it seems to me that all the detectives of fact and of fancy would be children in your hands. That's your line of life, sir, and you may take the word of a man who has seen something of the world."

'And that recommendation, with the exaggerated estimate of my ability with which he prefaced it, was, if you will believe me, Watson, the very first thing which ever made me feel that a profession might be made out of what had up to that time been the merest hobby. At the moment, however, I was too much concerned at the sudden illness of my host to think of anything else.

' "I hope that I have said nothing to pain you," said I.

' "Well, you certainly touched upon rather a tender point. Might I ask how you know and how much you know?" He spoke now in a half jesting fashion, but a look of terror still lurked at the back of his eyes.

' "It is simplicity itself," said I. "When you bared your arm to draw that fish into the boat, I saw that 'J. A.' had been tattooed in the bend of the elbow. The letters were still legible, but it was perfectly clear from their blurred appearance, and from the staining of the skin round them, that efforts had been made to obliterate them. It was obvious, then, that those initials had once been very familiar to you, and that you had afterwards wished to forget them."

' "What an eye you have!" he cried, with a sigh of relief. "It is just as you say. But we won't talk of it. Of all ghosts, the ghosts of our old loves are the worst. Come into the billiard-room and have a quiet cigar."

'From that day, amid all his cordiality, there was always a touch of suspicion in Mr Trevor's manner towards me. Even his son remarked it. "You've given the governor* such a turn," said he, "that he'll never be sure again of what you know and what you don't know." He did not mean to show it, I am sure, but it was so strongly in his mind that it peeped out at every action. At last I became so convinced that I was causing him uneasiness, that I drew my visit to a close. On the very day, however, before I left, an incident occurred which proved in the sequel to be of importance.

'We were sitting out upon the lawn on garden chairs, the three of us, basking in the sun and admiring the view across the Broads, when the maid came out to say that there was a man at the door who wanted to see Mr Trevor.

' "What is his name?" asked my host.

' "He would not give any."

' "What does he want, then?"

' "He says that you know him, and that he only wants a moment's conversation."

' "Show him round here." An instant afterwards there appeared a little wizened fellow, with a cringing manner and a shambling style of walking. He wore an open jacket, with a splotch of tar on the sleeve, a red and black check shirt, dungaree trousers, and heavy boots badly worn. His face was thin and brown and crafty, with a perpetual smile upon it, which showed an irregular line of yellow teeth, and his crinkled hands were half-closed in a way that is distinctive of sailors. As he came slouching across the lawn I heard Mr Trevor make a sort of hiccoughing noise in his throat, and, jumping out of his chair, he ran into the house. He was back in a moment, and I smelt a strong reek of brandy as he passed me.

' "Well, my man," said he, "what can I do for you?"

'The sailor stood looking at him with puckered eyes, and with the same loose-lipped smile upon his face.

' "You don't know me?" he asked.

' "Why, dear me, it is surely Hudson!" said Mr Trevor, in a tone of surprise.

' "Hudson it is, sir," said the seaman. "Why, it's thirty year and more since I saw you last. Here you are in your house, and me still picking my salt meat out of the harness cask."*

' "Tut, you will find that I have not forgotten old times," cried Mr Trevor, and, walking towards the sailor, he said something in a low voice. "Go into the kitchen," he continued out loud, "and you will get food and drink. I have no doubt that I shall find you a situation."

' "Thank you, sir," said the seaman, touching his forelock. "I'm just off a two-yearer in an eight-knot tramp,* short-handed at that, and I wants a rest. I thought I'd get it either with Mr Beddoes or with you."

' "Ah!" cried Mr Trevor, "you know where Mr Beddoes is?"

' "Bless you, sir, I know where all my old friends are," said the fellow, with a sinister smile, and slouched off after the maid to the kitchen. Mr Trevor mumbled something to us about having been shipmates with the man when he was going back to the diggings, and then, leaving us on the lawn, he went indoors. An hour later, when we entered the house we found him stretched dead drunk upon the dining-room sofa. The whole incident left a most ugly impression upon my mind, and I was not sorry next day to leave Donnithorpe behind me, for I felt that my presence must be a source of embarrassment to my friend.

'All this occurred during the first month of the long vacation. I went up to my London rooms, where I spent seven weeks working out a few experiments in organic chemistry. One day, however, when the autumn was far advanced and the vacation drawing to a close, I received a telegram from my friend imploring me to return to Donnithorpe, and saying that he was in great need of my advice and assistance. Of course I dropped everything and set out for the north once more.

'He met me with the dog-cart* at the station, and I saw at a glance that the last two months had been very trying ones for him. He had grown thin and careworn, and had lost the loud, cheery manner for which he had been remarkable.

' "The governor is dying," were the first words he said.

' "Impossible!" I cried. "What is the matter?"

' "Apoplexy.* Nervous shock. He's been on the verge all day. I doubt if we shall find him alive."

'I was, as you may think, Watson, horrified at this unexpected news.

' "What has caused it?" I asked.

' "Ah, that is the point. Jump in, and we can talk it over while we drive. You remember that fellow who came upon the evening before you left us?"

' "Perfectly."

' "Do you know who it was that we let into the house that day?"

' "I have no idea."

' "It was the Devil, Holmes!" he cried.

'I stared at him in astonishment.

' "Yes; it was the Devil himself. We have not had a peaceful hour since—not one. The governor has never held up his head from that evening, and now the life has been crushed out of him, and his heart broken, all through this accursed Hudson."

' "What power had he, then?"

' "Ah, that is what I would give so much to know. The kindly, charitable, good old governor! How could he have fallen into the clutches of such a ruffian? But I am so glad that you have come, Holmes. I trust very much to your judgment and discretion, and I know that you will advise me for the best."

'We were dashing along the smooth, white country road, with the long stretch of Broads* in front of us glimmering in the red light of the setting sun. From a grove upon our left I could already see the high chimneys and the flag-staff which marked the squire's dwelling.

' "My father made the fellow gardener," said my companion, "and then, as that did not satisfy him, he was promoted to be butler. The house seemed to be at his mercy, and he wandered about and did what he chose in it. The maids complained of his drunken habits and his vile language. The dad raised their wages all round to recompense them for the annoyance. The fellow would take the boat and my father's best gun and treat himself to little shooting parties. And all this with such a sneering, leering, insolent face, that I would have knocked him down twenty times over if he had been a man of my own age. I tell you, Holmes, I have had to keep a tight hold upon myself all this time, and now I am asking myself whether, if I had let myself go a little more, I might not have been a wiser man.

' "Well, matters went from bad to worse with us, and this animal, Hudson, became more and more intrusive, until at last, on his making some insolent reply to my father in my presence one day, I took him by the shoulder and turned him out of the room. He slunk away with a livid face, and two venomous eyes which uttered more threats than his tongue could do. I don't know what passed between the poor dad and him after that, but the dad came to me next day and asked me whether I would mind apologizing to Hudson. I refused, as you can imagine, and asked my father how he could allow such a wretch to take such liberties with himself and his household.

' "Ah, my boy," said he, "it is all very well to talk, but you don't know how I am placed. But you shall know, Victor. I'll see that you shall know, come what may! You wouldn't believe harm of your poor old father, would you, lad?" He was very much moved, and shut himself up in the study all day, where I could see through the window that he was writing busily.

' "That evening there came what seemed to me to be a grand release, for Hudson told us that he was going to leave us. He walked into the dining-room as we sat after dinner and announced his intention in the thick voice of a half-drunken man.

' "I've had enough of Norfolk," said he, "I'll run down to Mr Beddoes, in Hampshire. He'll be as glad to see me as you were, I dare say."

' "You're not going away in an unkind spirit, Hudson, I hope?" said my father, with a tameness which made my blood boil.

' "I've not had my 'pology," said he, sulkily, glancing in my direction.

' "Victor, you will acknowledge that you have used this worthy fellow rather roughly?" said the dad, turning to me.

' "On the contrary, I think that we have both shown extraordinary patience towards him," I answered.

' "Oh, you do, do you?" he snarled. "Very good, mate. We'll see about that!" He slouched out of the room, and half an hour afterwards left the house, leaving my father in a state of pitiable nervousness. Night after night I heard him pacing his room, and it was just as he was recovering his confidence that the blow did at last fall.

' "And how?" I asked eagerly.

' "In a most extraordinary fashion. A letter arrived for my father yesterday evening, bearing the Fordingbridge* post-mark. My father read it, clapped both his hands to his head, and began running round the room in little circles like a man who has been driven out of his senses. When I at last drew him down on to the sofa, his mouth and eyelids were all puckered on one side, and I saw that he had a stroke. Dr Fordham came over at once, and we put him to bed; but the paralysis has spread, he has shown no sign of returning consciousness, and I think that we shall hardly find him alive."

' "You horrify me, Trevor!" I cried. "What, then, could have been in this letter to cause so dreadful a result?"

' "Nothing. There lies the inexplicable part of it. The message was absurd and trivial. Ah, my God, it is as I feared!"

'As he spoke we came round the curve of the avenue, and saw in the fading light that every blind in the house had been drawn down. As we dashed up to the door, my friend's face convulsed with grief, a gentleman in black emerged from it.

' "When did it happen, doctor?" asked Trevor.

' "Almost immediately after you left."

' "Did he recover consciousness?"

' "For an instant before the end."

' "Any message for me?"

' "Only that the papers were in the back drawer of the Japanese cabinet.*"

'My friend ascended with the doctor to the chamber of death, while I remained in the study, turning the whole matter over and over in my head, and feeling as sombre as ever I had done in my life. What was the past of this Trevor: pugilist, traveller, and gold-digger; and how had he placed himself in the power of this acid-faced seaman? Why, too, should he faint at an allusion to the half-effaced initials upon his arm, and die of fright when he had a letter from Fordingbridge? Then I remembered that Fordingbridge was in Hampshire, and that this Mr Beddoes, whom the seaman had gone to visit, and presumably to blackmail, had also been mentioned as living in Hampshire. The letter, then, might either come from Hudson, the seaman, saying that he had betrayed the guilty secret which appeared to exist, or it might come from Beddoes, warning an old confederate that such a betrayal was imminent. So far it seemed clear enough. But, then, how could the letter be trivial and grotesque as described by the son? He must have misread it. If so, it must have been one of those ingenious secret codes which mean one thing while they seem to mean another. I must see this letter. If there were a hidden meaning in it, I was confident that I could pluck it forth. For an hour I sat pondering over it in the gloom, until at last a weeping maid brought in a lamp, and close at her heels came my friend Trevor, pale but composed, with these very papers, which lie upon my knee, held in his grasp. He sat down opposite to me, drew the lamp to the edge of the table, and handed me a short note scribbled, as you see, upon a single sheet of grey paper. "The supply of game for London is going steadily up," it ran. "Head-keeper Hudson, we believe, has been now told to receive all orders for fly paper and for preservation of your hen pheasant's life."

'I dare say my face looked as bewildered as yours did just now when first I read this message. Then I re-read it very carefully. It was evidently as I had thought, and some second meaning must be buried in this strange combination of words. Or could it be that there was a prearranged significance to such phrases as "fly paper" and "hen pheasant"? Such a meaning would be arbitrary, and could not be deduced in any way. And yet I was loth to believe that this was the case, and the presence of the word "Hudson" seemed to show that the subject of the message was as I had guessed, and that it was from Beddoes rather than the sailor. I tried it backwards, but the combination, "Life pheasant's hen," was not encouraging. Then I tried alternate words, but neither "The of for" nor "supply game London" promised to throw any light upon it. And then in an instant the key of the riddle was in my hands, and I saw that every third word beginning with the first would give a message which might well drive old Trevor to despair.

'It was short and terse, the warning, as I now read it to my companion:

' "The game is up. Hudson has told all. Fly for your life."

'Victor Trevor sank his face into his shaking hands. "It must be that, I suppose," said he. "This is worse than death, for it means disgrace as well. But what is the meaning of these 'head-keepers' and 'hen-pheasants'?"

' "It means nothing to the message, but it might mean a good deal to us if we had no other means of discovering the sender. You see that he has begun by writing, 'The . . . game . . . is,' and so on. Afterwards he had, to fulfil the prearranged cipher, to fill in any two words in each space. He would naturally use the first words which came to his mind, and if there were so many which referred to sport among them, you may be tolerably sure that he is either an ardent shot or interested in breeding. Do you know anything of this Beddoes?"

' "Why, now that you mention it," said he, "I remember that my poor father used to have an invitation from him to shoot over his preserves every autumn."

' "Then it is undoubtedly from him that the note comes," said I. "It only remains for us to find out what this secret was which the sailor Hudson seems to have held over the heads of these two wealthy and respected men."

' "Alas, Holmes, I fear that it is one of sin and shame!" cried my friend. "But from you I shall have no secrets. Here is the statement which was drawn up by my father when he knew that the danger from Hudson had become imminent. I found it in the Japanese cabinet, as he told the doctor. Take it and read it to me, for I have neither the strength nor the courage to do it myself."

'These are the very papers, Watson, which he handed to me, and I will read them to you as I read them in the old study that night to him. They are indorsed outside, as you see: "Some particulars of the voyage of the barque* *Gloria Scott*, from her leaving Falmouth on the 8th October, 1855, to her destruction in N. lat. 15° 20', W. long. 25° 14', on November 6th." It is in the form of a letter, and runs in this way:

'My dear, dear son,—Now that approaching disgrace begins to darken the closing years of my life, I can write with all truth and honesty that it is not the terror of the law, it is not the loss of my position in the county, nor is it my fall in the eyes of all who have known me, which cuts me to the heart; but it is the thought that you should come to blush for me—you who love me, and who have seldom, I hope, had reason to do other than respect me. But if the blow falls which is for ever hanging over me, then I should wish you to read this that you may know straight from me how far I have been to blame. On the other hand, if all should go well (which may kind God Almighty grant!), then if by any chance this paper should be still undestroyed, and should fall into your hands, I conjure you by all you hold sacred, by the memory of your dear mother, and by the love which has been between us, to hurl it into the fire, and to never give one thought to it again.

'If, then, your eye goes on to read this line, I know that I shall already have been exposed and dragged from my home, or, as is more likely—for you know that my heart is weak—be lying with my tongue sealed for ever in death. In either case the time for suppression is past, and every word which I tell you is the naked truth; and this I swear as I hope for mercy.

'My name, dear lad, is not Trevor. I was James Armitage in my younger days, and you can understand now the shock that it was to me a few weeks ago when your college friend addressed me in words which seemed to imply that he had surmised my secret. As Armitage it was that I entered a London banking house, and as Armitage I was convicted of breaking my country's laws, and was sentenced to transportation.* Do not think very harshly of me, laddie. It was a debt of honour, so-called, which I had to pay, and I used money which was not my own to do it, in the certainty that I could replace it before there could be any possibility of its being missed. But the most dreadful ill-luck pursued me. The money which I had reckoned upon never came to hand, and a premature examination of accounts exposed my deficit. The case might have been dealt leniently with, but the laws were more harshly administered thirty years ago than now, and on my twenty-third birthday I found myself chained as a felon with thirty-seven other convicts in the 'tween decks of the barque *Gloria Scott*, bound for Australia.

'It was the year '55, when the Crimean War was at its height,* and the old convict ships had been largely used as transports in the Black Sea. The Government was compelled therefore to use smaller and less suitable vessels for sending out their prisoners. The *Gloria Scott* had been in the Chinese tea trade, but she was an old-fashioned, heavy-bowed, broad-beamed craft, and the new clippers* had cut her out. She was a 500-ton boat, and besides her thirty-eight jail-birds, she carried twenty-six of a crew, eighteen soldiers, a captain, three mates, a doctor, a chaplain, and four warders. Nearly a hundred souls were in her, all told, when we set sail from Falmouth.

'The partitions between the cells of the convicts, instead of being of thick oak, as is usual in convict ships, were quite thin and frail. The man next to me upon the aft side was one whom I had particularly noticed when we were led down to the quay. He was a young man with a clear, hairless face, a long thin nose, and rather nutcracker jaws. He carried his head very jauntily in the air, had a swaggering style of walking, and was above all else remarkable for his extraordinary height. I don't think any of our heads would come up to his shoulder, and I am sure that he could not have measured less than six and a half feet. It was strange among so many sad and weary faces to see one which was full of energy and resolution. The sight of it was to me like a fire in a snowstorm. I was glad then to find that he was my neighbour, and gladder still when, in the dead of the night, I heard a whisper close to my ear, and found that he had managed to cut an opening in the board which separated us.

' "Hullo, chummy!" said he, "what's your name, and what are you here for?"

'I answered him, and asked in turn who I was talking with.

' "I'm Jack Prendergast," said he, "and, by God, you'll learn to bless my name before you've done with me!"

'I remembered hearing of his case, for it was one which had made an immense sensation throughout the country, some time before my own arrest. He was a man of good family and of great ability, but of incurably vicious habits, who had, by an ingenious system of fraud, obtained huge sums of money from the leading London merchants.

' "Ah, ah! You remember my case?" said he, proudly.

' "Very well indeed."

' "Then maybe you remember something queer about it?"

' "What was that, then?"

' "I had nearly a quarter of a million, hadn't I?"

' "So it was said."

' "But none was recovered, eh?"

' "No."

' "Well, where d'ye suppose the balance is?" he asked.

' "I have no idea," said I.

' "Right between my finger and thumb," he cried. "By God, I've got more pounds to my name than you have hairs on your head. And if you've money, my son, and know how to handle it and spread it, you can do *anything*! Now, you don't think it likely that a man who could do anything is going to wear his breeches out sitting in the stinking hold of a rat-gutted, beetle-ridden, mouldy old coffin of a China coaster? No, sir, such a man will look after himself, and will look after his chums. You may lay to that! You hold on to him, and you may kiss the Book* that he'll haul you through."

'That was his style of talk, and at first I thought it meant nothing, but after a while, when he had tested me and sworn* me in with all possible solemnity, he let me understand that there really was a plot to gain command of the vessel. A dozen of the prisoners had hatched it before they came aboard; Prendergast was the leader, and his money was the motive power.

' "I'd a partner," said he, "a rare good man, as true as a stock to a barrel. He's got the dibbs,* he has, and where do you think he is at this moment? Why, he's the chaplain of this ship—the chaplain, no less! He came aboard with a black coat and his papers right, and money enough in his box to buy the thing right from keel to main-truck.* The crew are his, body and soul. He could buy 'em at so much a gross with a cash discount, and he did it before ever they signed on. He's got two of the warders and Mercer the second mate, and he'd get the captain himself if he thought him worth it."

' "What are we to do, then?" I asked.

' "What do you think?" said he. "We'll make the coats of some of these soldiers redder than ever the tailor did."

' "But they are armed," said I.

' "And so shall we be, my boy. There's a brace of pistols for every mother's son of us, and if we can't carry this ship, with the crew at our back, it's time we were all sent to a young Miss's boarding school. You speak to your mate on the left to-night, and see if he is to be trusted."

'I did so, and found my other neighbour to be a young fellow in much the same position as myself, whose crime had been forgery. His name was Evans but he afterwards changed it, like myself, and he is now a rich and prosperous man in the South of England. He was ready enough to join the conspiracy, as the only means of saving ourselves, and before we had crossed the Bay* there were only two of the prisoners who were not in the secret. One of these was of weak mind, and we did not dare to trust him, and the other was suffering from jaundice, and could not be of any use to us.

'From the beginning there was really nothing to prevent us taking possession of the ship. The crew were a set of ruffians, specially picked for the job. The sham chaplain came into our cells to exhort us, carrying a black bag, supposed to be full of tracts;* and so often did he come that by the third day we had each stowed away at the foot of our bed a file, a brace of pistols, a pound of powder, and twenty slugs. Two of the warders were agents of Prendergast, and the second mate was his right-hand man. The captain, the two mates, two warders, Lieutenant Martin, his eighteen soldiers, and the doctor were all that we had against us. Yet, safe as it was, we determined to neglect no precaution, and to make our attack suddenly at night. It came, however, more quickly than we expected, and in this way:

'One evening, about the third week after our start, the doctor had come down to see one of the prisoners, who was ill, and, putting his hand down the bottom of his bunk, he felt the outline of the pistols. If he had been silent he might have blown the whole thing; but he was a nervous little chap, so he gave a cry of surprise and turned so pale, that the man knew what was up in an instant and seized him. He was gagged before he could give the alarm, and tied down upon the bed. He had unlocked the door that led to the deck, and we were through it in a rush. The two sentries were shot down, and so was a corporal who came running to see what was the matter. There were two more soldiers at the door of the state-room, and their muskets seemed not to be loaded, for they never fired upon us, and they were shot while trying to fix their bayonets. Then we rushed on into the captain's cabin, but as we pushed open the door there was an explosion from within, and there he lay with his head on the chart of the Atlantic,* which was pinned upon the table, while the chaplain stood, with a smoking pistol in his hand, at his elbow. The two mates had both been seized by the crew, and the whole business seemed to be settled.

'The state-room was next the cabin, and we flocked in there and flopped down on the settees, all speaking together, for we were just mad with the feeling that we were free once more. There were lockers all round, and Wilson, the sham chaplain, knocked one of them in, and pulled out a dozen of brown sherry. We cracked off the necks of the bottles, poured the stuff out into tumblers, and were just tossing them off, when in an instant, without warning, there came the roar of muskets in our ears, and the saloon was so full of smoke that we could not see across the table. When it cleared away again the place was a shambles. Wilson and eight others were wriggling on the top of each other on the floor, and the blood and the brown sherry on that table turn me sick now when I think of it. We were so cowed by the sight that I think we should have given the job up if it had not been for Prendergast. He bellowed like a bull and rushed for the door with all that were left alive at his heels. Out we ran, and there on the poop were the lieutenant and ten of his men. The swing skylights above the saloon table had been a bit open, and they had fired on us through the slit. We got on them before they could load, and they stood to it like men, but we had the upper hand of them, and in five minutes it was all over. My God! was there ever a slaughter-house like that ship? Prendergast was like a raging devil, and he picked the soldiers up as if they had been children and threw them overboard, alive or dead. There was one ser-geant that was horribly wounded, and yet kept on swim-ming for a surprising time, until someone in mercy blew out his brains. When the fighting was over there was no one left of our enemies except just the warders, the mates, and the doctor.

'It was over them that the great quarrel arose. There were many of us who were glad enough to win back our freedom, and yet who had no wish to have murder on our souls. It was one thing to knock the soldiers over with their muskets in their hands, and it was another to stand by while men were being killed in cold blood. Eight of us, five convicts and three sailors, said that we would not see it done. But there was no moving Prendergast and those who were with him. Our only chance of safety lay in making a clean job of it, said he, and he would not leave a tongue with power to wag in a witness-box. It nearly came to our sharing the fate of the prisoners, but at last he said that if we wished we might take a boat and go. We jumped at the offer, for we were already sick of these bloodthirsty doings, and we saw that there would be worse before it was done. We were given a suit of sailors' togs each, a barrel of water, two casks, one of junk* and one of biscuits, and a compass. Prendergast threw us over a chart, told us that we were shipwrecked mariners whose ship had foundered in lat. 15° N. and long. 25° W., and then cut the painter* and let us go.

'And now I come to the most surprising part of my story, my dear son. The seamen had hauled the foreyard aback* during the rising, but now as we left them they brought it square again, and, as there was a light wind from the north and east, the barque began to draw slowly away from us. Our boat lay, rising and falling, upon the long, smooth rollers, and Evans and I, who were the most educated of the party, were sitting in the sheets working out our position and planning what coast we should make for. It was a nice question, for the Cape de Verds* was about 500 miles to the north of us, and the African coast about 700 miles to the east. On the whole, as the wind was coming round to north, we thought that Sierra Leone might be best, and turned our head in that direction, the barque being at that time nearly hull down on our starboard quarter. Suddenly as we looked at her we saw a dense black cloud of smoke shoot up from her, which hung like a monstrous tree upon the sky-line. A few seconds later a roar like thunder burst upon our ears, and as the smoke thinned away there was no sign left of the *Gloria Scott*. In an instant we swept the boat's head round again, and pulled with all our strength for the place where the haze, still trailing over the water, marked the scene of this catastrophe.

'It was a long hour before we reached it, and at first we feared that we had come too late to save anyone. A splintered boat and a number of crates and fragments of spars rising and falling on the waves showed us where the vessel had foundered, but there was no sign of life, and we had turned away in despair when we heard a cry for help, and saw at some distance a piece of wreckage with a man lying stretched across it. When we pulled him aboard the boat he proved to be a young seaman of the name of Hudson, who was so burned and exhausted that he could give us no account of what had happened until the following morning.

'It seemed that, after we had left, Prendergast and his gang had proceeded to put to death the remaining five prisoners:

the two warders had been shot and thrown overboard, and so also had the third mate. Prendergast then descended into the 'tween decks, and with his own hands cut the throat of the unfortunate surgeon. There only remained the first mate, who was a bold and active man. When he saw the convict approaching him with the bloody knife in his hand, he kicked off his bonds, which he had somehow contrived to loosen, and rushing down the deck he plunged into the after-hold.

'A dozen convicts who descended with their pistols in search of him found him with a match-box in his hand seated beside an open powder barrel, which was one of a hundred carried on board, and swearing that he would blow all hands up if he were in any way molested. An instant later the explosion occurred, though Hudson thought it was caused by the misdirected bullet of one of the convicts rather than the mate's match. Be the cause what it may, it was the end of the *Gloria Scott*, and of the rabble who held command of her.

'Such, in a few words, my dear boy, is the history of this terrible business in which I was involved. Next day we were picked up by the brig *Hotspur*, bound for Australia, whose captain found no difficulty in believing that we were the survivors of a passenger ship which had foundered. The transport ship, *Gloria Scott*, was set down by the Admiralty as being lost at sea, and no word has ever leaked out as to her true fate. After an excellent voyage the *Hotspur* landed us at Sydney, where Evans and I changed our names and made our way to the diggings, where among the crowds who were gathered from all nations, we had no difficulty in losing our former identities.

'The rest I need not relate. We prospered, we travelled, we came back as rich Colonials to England, and we bought country estates. For more than twenty years we have led peaceful and useful lives, and we hoped that our past was for ever buried. Imagine, then, my feelings when in the seaman who came to us I recognized instantly the man who had been picked off the wreck! He had tracked us down somehow, and had set himself to live upon our fears. You will understand now how it was that I strove to keep peace with him, and you will in some measure sympathize with me in the fears which fill me, now that he has gone from me to his other victim with threats upon his tongue.

'Underneath is written, in a hand so shaky as to be hardly legible, "Beddoes writes in cipher to say that H. has told all. Sweet Lord, have mercy on our souls!"

'That was the narrative which I read that night to young Trevor, and I think, Watson, that under the circumstances it was a dramatic one. The good fellow was heartbroken at it, and went out to the Terai* tea planting, where I hear that he is doing well. As to the sailor and Beddoes, neither of them was ever heard of again after that day on which the letter of warning was written. They both disappeared utterly and completely. No complaint had been lodged with the police, so that Beddoes had mistaken a threat for a deed. Hudson had been seen lurking about, and it was believed by the police that he had done away with Beddoes, and had fled. For myself, I believe that the truth was exactly the opposite. I think it is most probable that Beddoes, pushed to desperation, and believing himself to have been already betrayed, had revenged himself upon Hudson, and had fled from the country with as much money as he could lay his hands on. Those are the facts of the case, Doctor, and if they are of any use to your collection, I am sure that they are very heartily at your service.'

The Musgrave Ritual

A N anomaly which often struck me in the character of my friend Sherlock Holmes was that, although in his methods of thought he was the neatest and most methodical of mankind, and although also he affected a certain quiet primness of dress, he was none the less in his personal habits one of the most untidy men that ever drove a fellow-lodger to distraction. Not that I am in the least conventional in that respect myself. The rough-and-tumble work in Afghanistan,* coming on the top of a natural Bohemianism of disposition, has made me rather more lax than befits a medical man. But with me there is a limit, and when I find a man who keeps his cigars in the coal-scuttle, his tobacco in the toe end of a Persian slipper, and his unanswered correspondence transfixed by a jack-knife into the very centre of his wooden mantelpiece, then I begin to give myself virtuous airs. I have always held, too, that pistol practice should distinctly be an open-air pastime; and when Holmes in one of his queer humours would sit in an arm-chair, with his hair-trigger* and a hundred Boxer cartridges,* and proceed to adorn the opposite wall with a patriotic V.R.* done in bullet-pocks, I felt strongly that neither the atmosphere nor the appearance of our room was improved by it.

Our chambers were always full of chemicals and of criminal relics, which had a way of wandering into unlikely positions, and of turning up in the butter-dish, or in even less desirable places. But his papers were my great crux. He had a horror of destroying documents, especially those which were connected with his past cases, and yet it was only once in every year or two that he would muster energy to docket and arrange them, for, as I have mentioned somewhere in these incoherent memoirs,* the outbursts of passionate energy when he performed the remarkable feats with which his name is associated were followed by reactions of lethargy, during which he would lie about with his violin and his books, hardly moving, save from the sofa to the table. Thus month after month his papers accumulated, until every corner of the room was stacked with bundles of manuscript which were on no account to be burned, and which could not be put away save by their owner.

One winter's night, as we sat together by the fire, I ventured to suggest to him that as he had finished pasting extracts into his commonplace book he might employ the next two hours in making our room a little more habitable. He could not deny the justice of my request, so with a rather rueful face he went off to his bedroom, from which he returned presently pulling a large tin box behind him. This he placed in the middle of the floor, and squatting down upon a stool in front of it he threw back the lid. I could see that it was already a third full of bundles of paper tied up with red tape into separate packages.

'There are cases enough here, Watson,' said he, looking at me with mischievous eyes. 'I think that if you knew all that I have* in this box you would ask me to pull some out instead of putting others in.'

'These are the records of your early work, then?' I asked. 'I have often wished that I had notes of those cases.'

'Yes, my boy; these were all done prematurely before my biographer had come to glorify me.' He lifted bundle after bundle in a tender, caressing sort of way. 'They are not all successes, Watson,' said he, 'but there are some pretty little problems among them. Here's the record of the Tarleton* murders and the case of Vamberry, the wine merchant, and the adventure of the old Russian woman, and the singular affair of the aluminium crutch, as well as a full account of Ricoletti of the club foot* and his abominable wife. And here—ah, now! this really is something a little *recherché*.'*

He dived his arm down to the bottom of the chest, and brought up a small wooden box, with a sliding lid, such as children's toys are kept in. From within he produced a crumpled piece of paper, an old-fashioned brass key, a peg of wood with a ball of string attached to it, and three rusty old discs of metal.

'Well, my boy, what do you make of this lot?' he asked, smiling at my expression.

'It is a curious collection.'

'Very curious; and the story that hangs round it will strike you as being more curious still.'

'These relics have a history, then?'

'So much so that they *are* history.'

'What do you mean by that?'

Sherlock Holmes picked them up one by one, and laid them along the edge of the table. Then he re-seated himself in his chair, and looked them over with a gleam of satisfaction in his eyes.

'These,' said he, 'are all that I have left to remind me of the episode of the Musgrave Ritual.'*

I had heard him mention the case more than once, though I had never been able to gather the details.

'I should be so glad,' said I, 'if you would give me an account of it.'

'And leave the litter as it is?' he cried, mischievously. 'Your tidiness won't bear much strain, after all, Watson. But I should be glad that you should add this case to your annals, for there are points in it which make it quite unique in the criminal records of this or, I believe, of any other country. A collection of my trifling achievements would certainly be incomplete which contained no account of this very singular business.

'You may remember how the affair of the *Gloria Scott*, and my conversation with the unhappy man whose fate I told you of, first turned my attention in the direction of the profession which has become my life's work. You see me now when my name has become known far and wide, and when I am generally recognised both by the public and by the official force as being a final court of appeal in doubtful cases. Even when you knew me first, at the time of the affair which you have commemorated in *A Study of Scarlet*, I had already established a considerable, though not a very lucrative, connection. You can hardly realize, then, how difficult I found it at first, and how long I had to wait before I succeeded in making any headway.

'When I first came up to London I had rooms in Montague Street,* just round the corner from the British Museum, and there I waited, filling in my too abundant leisure time by studying all those branches of science which might make me more efficient. Now and again cases came in my way, principally through the introduction of old fellow students, for during my last years at the university there was a good deal of talk there about myself and my methods. The third of these cases was that of the Musgrave Ritual, and it is to the interest which was aroused by that singular chain of events, and the large issues which proved to be at stake, that I trace my first stride towards the position which I now hold.

'Reginald Musgrave had been in the same college as myself, and I had some slight acquaintance with him. He was not generally popular among the undergraduates, though it always seemed to me that what was set down as pride was really an attempt to cover extreme natural diffidence. In appearance he was a man of an exceedingly aristocratic type, thin, high-nosed, and large-eyed, with languid and yet courtly manners. He was indeed a scion of one of the very oldest families in the kingdom, though his branch was a cadet one which had separated from the northern Musgraves some time in the sixteenth century, and had established itself in western Sussex, where the manor house of Hurlstone is perhaps the oldest inhabited building in the county. Something of his birthplace seemed to cling to the man, and I never looked at his pale, keen face, or the poise of his head, without associating him with grey archways and mullioned windows* and all the venerable wreckage of a feudal keep.* Now and again we drifted into talk, and I can remember that more than once he expressed a keen interest in my methods of observation and inference.

'For four years I had seen nothing of him, until one morning he walked into my room in Montague Street. He had changed little, was dressed like a young man of fashion—he was always a bit of a dandy—and preserved the same quiet, suave manner which had formerly distinguished him.

' "How has all gone with you, Musgrave?" I asked, after we had cordially shaken hands.

' "You probably heard of my poor father's death," said he. "He was carried off about two years ago. Since then I have, of course, had the Hurlstone estates to manage, and as I am member* for the district as well, my life has been a busy one; but I understand, Holmes, that you are turning to practical ends those powers with which you used to amaze us."

' "Yes," said I, "I have taken to living by my wits."

' "I am delighted to hear it, for your advice at present would be exceedingly valuable to me. We have had some very strange doings at Hurlstone, and the police have been able to throw no light upon the matter. It is really the most extraordinary and inexplicable business."

'You can imagine with what eagerness I listened to him, Watson, for the very chance for which I had been panting during all those months of inaction seemed to have come within my reach. In my inmost heart I believed that I could succeed where others failed, and now I had the opportunity to test myself.

' "Pray let me have the details," I cried.

'Reginald Musgrave sat down opposite to me, and lit the cigarette which I had pushed towards him.

' "You must know," said he, "that though I am a bachelor I have to keep up a considerable staff of servants at Hurlstone, for it is a rambling old place, and takes a good deal of looking after. I preserve,* too, and in the pheasant months* I usually have a house party, so that it would not do to be short-handed. Altogether there are eight maids, the cook, the butler, two footmen, and a boy. The garden and the stables, of course, have a separate staff.

' "Of these servants the one who had been longest in our service was Brunton, the butler. He was a young schoolmaster out of place when he was first taken up by my father, but he was a man of great energy and character, and he soon became quite invaluable in the household. He was a well-grown, handsome man, with a splendid forehead, and though he has been with us for twenty years he cannot be more than forty now. With his personal advantages and his extraordinary gifts, for he can speak several languages and play nearly every musical instrument, it is wonderful that he should have been satisfied so long in such a position, but I suppose that he was comfortable and lacked energy to make any change. The butler of Hurlstone is always a thing that is remembered by all who visit us.

' "But this paragon has one fault. He is a bit of a Don Juan,* and you can imagine that for a man like him it is not a very difficult part to play in a quiet country district.

' "When he was married it was all right, but since he has been a widower we* have had no end of trouble with him. A few months ago we were in hopes that he was about to settle down again, for he became engaged to Rachel Howells, our second housemaid, but he has thrown her over since then and taken up with Janet Tregellis, the daughter of the head gamekeeper. Rachel, who is a very good girl, but of an excitable Welsh temperament, had a sharp touch of brain fever, and goes about the house now—or did until yesterday—like a black-eyed shadow of her former self. That was our first drama at Hurlstone, but a second one came to drive it from our minds, and it was prefaced by the disgrace and dismissal of butler Brunton.

' "This is how it came about. I have said that the man was intelligent, and this very intelligence has caused his ruin, for it seems to have led to an insatiable curiosity about things which did not in the least concern him. I had no idea of the lengths to which this would carry him until the merest accident opened my eyes to it.

' "I have said that the house is a rambling one. One night last week—on Thursday night, to be more exact—I found that I could not sleep, having foolishly taken a cup of strong *café noir** after my dinner. After struggling against it until two in the morning I felt that it was quite hopeless, so I rose and lit the candle with the intention of continuing a novel which I was reading. The book, however, had been left in the billiard-room, so I pulled on my dressing-gown and started off to get it.

' "In order to reach the billiard-room I had to descend a flight of stairs, and then to cross the head of the passage which led to the library and the gun-room. You can imagine my surprise when as I looked down this corridor I saw a glimmer of light coming from the open door of the library. I had myself extinguished the lamp and closed the door before coming to bed. Naturally, my first thought was of burglars. The corridors at Hurlstone have their walls largely decorated with trophies of old weapons. From one of these I picked a battle-axe, and then, leaving my candle behind me, I crept on tip-toe down the passage and peeped in at the open door.

' "Brunton, the butler, was in the library. He was sitting, fully dressed, in an easy chair, with a slip of paper, which looked like a map, upon his knee, and his forehead sunk forward upon his hand in deep thought. I stood, dumb with astonishment, watching him from the darkness. A small taper on the edge of the table shed a feeble light, which sufficed to show me that he was fully dressed. Suddenly, as I looked, he rose from his chair, and walking over to a bureau at the side, he unlocked it and drew out one of the drawers. From this he took a paper, and, returning to his seat, he flattened it out beside the taper on the edge of the table, and began to study it with minute attention. My indignation at this calm examination of our family documents overcame me so far that I took a step forward, and Brunton looking up saw me standing in the doorway. He sprang to his feet, his face turned livid with fear, and he thrust into his breast the chart-like paper which he had been originally studying.

' " 'So!' said I, 'this is how you repay the trust which we have reposed in you! You will leave my service tomorrow.'

' "He bowed with the look of a man who is utterly crushed, and slunk past me without a word. The taper was still on the table, and by its light I glanced to see what the paper was which Brunton had taken from the bureau. To my surprise it was nothing of any importance at all, but simply a copy of the questions and answers in the singular old observance called the Musgrave Ritual. It is a sort of ceremony peculiar to our family, which each Musgrave for centuries past has gone through upon his coming of age—a thing of private interest, and perhaps of some little importance to the archaeologist, like our own blazonings and charges,* but of no practical use whatever."

' "We had better come back to the paper afterwards," said I.

' "If you think it really necessary," he answered, with some hesitation. "To continue my statement, however, I re-locked the bureau, using the key which Brunton had left, and I had turned to go, when I was surprised to find that the butler had returned and was standing before me.

' " 'Mr Musgrave, sir,' he cried, in a voice which was hoarse with emotion, 'I can't bear disgrace, sir. I've always been proud above my station in life, and disgrace would kill me. My blood will be on your head, sir—it will, indeed—if you drive me to despair. If you cannot keep me after what has passed, then for God's sake let me give you notice and leave in a month, as if of my own free will. I could stand that, Mr Musgrave, but not to be cast out before all the folk that I know so well.'

' " 'You don't deserve much consideration, Brunton,' I answered. 'Your conduct has been most infamous. However, as you have been a long time in the family, I have no wish to bring public disgrace upon you. A month, however, is too long. Take yourself away in a week, and give what reason you like for going.'

' " 'Only a week, sir?" he cried in a despairing voice. 'A fortnight—say at least a fortnight.'

' " 'A week,' I repeated, 'and you may consider yourself to have been very leniently dealt with.'

' "He crept away, his face sunk upon his breast, like a broken man, while I put out the light and returned to my room.

' "For two days after this Brunton was most assiduous in his attention to his duties. I made no allusion to what had passed, and waited with some curiosity to see how he would cover his disgrace. On the third morning, however, he did not appear, as was his custom, after breakfast to receive my instructions for the day. As I left the dining-room I happened to meet Rachel Howells, the maid. I have told you that she had only recently recovered from an illness, and was looking so wretchedly pale and wan that I remonstrated with her for being at work.

' " 'You should be in bed,' I said. 'Come back to your duties when you are stronger.'

' "She looked at me with so strange an expression that I began to suspect that her brain was affected.

' " 'I am strong enough, Mr Musgrave,' said she.

' " 'We will see what the doctor says,' I answered. 'You must stop work now, and when you go downstairs just say that I wish to see Brunton.'

' " 'The butler is gone,' said she.

' " 'Gone! Gone where?'

' " 'He is gone. No one has seen him. He is not in his room. Oh, yes, he is gone—he is gone!' She fell back against the wall with shriek after shriek of laughter, while I, horrified at this sudden hysterical attack, rushed to the bell to summon help. The girl was taken to her room, still screaming and sobbing, while I made inquiries about Brunton. There was no doubt about it that he had disappeared. His bed had not been slept in; he had been seen by no one since he had retired to his room the night before; and yet it was difficult to see how he could have left the house, as both windows and doors were found to be fastened in the morning. His clothes, his watch, and even his money were in his room—but the black suit which he usually wore was missing. His slippers, too, were gone, but his boots were left behind. Where, then, could butler Brunton have gone in the night, and what could have become of him now?

' "Of course we searched the house and the outhouses,* but there was no trace of him. It is, as I have said, a labyrinth of an old building,* especially the original wing, which is now practically uninhabited, but we ransacked every room and attic without discovering the least sign of the missing man. It was incredible to me that he could have gone away leaving all his property behind him, and yet where could he be? I called in the local police, but without success. Rain had fallen on the night before, and we examined the lawn and the paths all round the house, but in vain. Matters were in this state when a new development quite drew our attention away from the original mystery.

' "For two days Rachel Howells had been so ill, sometimes delirious, sometimes hysterical, that a nurse had been employed to sit up with her at night. On the third night after Brunton's disappearance, the nurse, finding her patient sleeping nicely, had dropped into a nap in the arm-chair, when she woke in the early morning to find the bed empty, the window open, and no signs of the invalid. I was instantly aroused, and with the two footmen started off at once in search of the missing girl. It was not difficult to tell the direction which she had taken, for, starting from under her window, we could follow her footmarks easily across the lawn to the edge of the mere,* where they vanished, close to the gravel path which leads out of the grounds. The lake there is 8 ft. deep, and you can imagine our feelings when we saw that the trail of the poor demented girl came to an end at the edge of it.

' "Of course, we had the drags at once, and set to work to recover the remains; but no trace of the body could we find. On the other hand, we brought to the surface an object of a most unexpected kind. It was a linen bag, which contained within it a mass of old rusted and discoloured metal and several dull-coloured pieces of pebble or glass. This strange find was all that we could get from the mere, and although we made every possible search and inquiry yesterday, we know nothing of the fate either of Rachel Howells or Richard Brunton. The county police are at their wits' end, and I have come up to you as a last resource."

'You can imagine, Watson, with what eagerness I listened to this extraordinary sequence of events, and endeavoured to piece them together, and to devise some common thread upon which they might all hang.

'The butler was gone. The maid was gone. The maid had loved the butler, but had afterwards had cause to hate him. She was of Welsh blood, fiery and passionate. She had been terribly excited immediately after his disappearance. She had flung into the lake a bag containing some curious contents. These were all factors which had to be taken into consideration, and yet none of them got quite to the heart of the matter. What was the starting-point of this chain of events? There lay the end of this tangled line.

' "I must see that paper, Musgrave," said I, "which this butler of yours thought it worth his while to consult, even at the risk of the loss of his place."

' "It is rather an absurd business, this Ritual of ours," he answered, "but it has at least the saving grace of antiquity to excuse it. I have a copy of the questions and answers here, if you care to run your eye over them."

'He handed me the very paper which I have here, Watson, and this is the strange catechism to which each Musgrave had to submit when he came to man's estate. I will read you the questions and answers as they stand:

' "Whose was it?

' "His who is gone.*

' "Who shall have it?

' "He who will come.

' "What was the month?

' "The sixth from the first.

' "Where was the sun?

' "Over the oak.

' "Where was the shadow?

' "Under the elm.

' "How was it stepped?

' "North by ten and by ten, east by five and by five, south by two and by two, west by one and by one, and so under.

' "What shall we give for it?

' "All that is ours.

' "Why should we give it?

' "For the sake of the trust."

' "The original has no date, but is in the spelling of the middle of the seventeenth century," remarked Musgrave. "I am afraid, however, that it can be of little help to you in solving this mystery."

' "At least," said I, "it gives us another mystery, and one which is even more interesting than the first. It may be that the solution of the one may prove to be the solution of the other. You will excuse me, Musgrave, if I say that your butler appears to me to have been a very clever man, and to have had a clearer insight than ten generations of his masters." *

' "I hardly follow you," said Musgrave. "The paper seems to me to be of no practical importance."

' "But to me it seems immensely practical, and I fancy that Brunton took the same view. He had probably seen it before that night on which you caught him."

' "It is very possible. We took no pains to hide it."

' "He simply wished, I should imagine, to refresh his memory upon that last occasion. He had, as I understand, some sort of map or chart which he was comparing with the manuscript, and which he thrust into his pocket when you appeared?"

' "That is true. But what could he have to do with this old family custom of ours, and what does this rigmarole mean?"

' "I don't think that we should have much difficulty in determining that," said I. "With your permission we will take the first train down to Sussex and go a little more deeply into the matter upon the spot."

'The same afternoon saw us both at Hurlstone. Possibly you have seen pictures and read descriptions of the famous old building, so I will confine my account of it to saying that it is built in the shape of an L, the long arm being the more modern portion, and the shorter the ancient nucleus from which the other has developed. Over the low, heavy-lintelled door, in the centre of this old part, is chiselled the date 1607, but experts are agreed that the beams and stonework are really much older than this. The enormously thick walls and tiny windows of this part had in the last century driven the family into building the new wing, and the old one was used now as a storehouse and a cellar when it was used at all. A splendid park, with fine old timber, surrounded the house, and the lake, to which my client had referred, lay close to the avenue, about two hundred yards from the building.

'I was already firmly convinced, Watson, that there were not three separate mysteries here, but one only, and that if I could read the Musgrave Ritual aright, I should hold in my hand the clue which would lead me to the truth concerning both the butler Brunton, and the maid Howells. To that, then, I turned all my energies. Why should this servant be so anxious to master this old formula? Evidently because he saw something in it which had escaped all those generations of country squires, and from which he expected some personal advantage. What was it, then, and how had it affected his fate?

'It was perfectly obvious to me on reading the Ritual that the measurements must refer to some spot to which the rest of the document alluded, and that if we could find that spot we should be in a fair way towards knowing what the secret was which the old Musgraves had thought it necessary to embalm in so curious a fashion. There were two guides given us to start with, an oak and an elm. As to the oak, there could be no question at all. Right in front of the house, upon the left-hand side of the drive, there stood a patriarch among oaks, one of the most magnificent trees that I have ever seen.

' "That was there when your Ritual was drawn up?" said I, as we drove past it.

' "It was there at the Norman Conquest, in all probability," he answered. "It has a girth of 23 ft."

'Here was one of my fixed points secured.

' "Have you any old elms?" I asked.

' "There used to be a very old one over yonder, but it was struck by lightning ten years ago, and we cut down the stump."

' "You can see where it used to be?"

' "Oh, yes."

' "There are no other elms?"

' "No old ones, but plenty of beeches."

' "I should like to see where it grew."

'We had driven up in a dog-cart, and my client led me away at once, without our entering the house, to the scar on the lawn where the elm had stood. It was nearly midway between the oak and the house. My investigation seemed to be progressing.

' "I suppose it is impossible to find out how high the elm was?" I asked.

' "I can give you it at once. It was 64 ft."

' "How do you come to know it?" I asked in surprise.

' "When my old tutor used to give me an exercise in trigonometry it always took the shape of measuring heights. When I was a lad I worked out every tree and building on the estate."

'This was an unexpected piece of luck. My data were coming more quickly than I could have reasonably hoped.

' "Tell me," I asked, "did your butler ever ask you such a question?"

'Reginald Musgrave looked at me in astonishment. "Now that you call it to my mind," he answered, "Brunton *did* ask me about the height of the tree some months ago, in connection with some little argument with the groom."

'This was excellent news, Watson, for it showed me that I was on the right road. I looked up at the sun. It was low in the heavens, and I calculated that in less than an hour it would lie just above the topmost branches of the old oak. One condition mentioned in the Ritual would then be fulfilled. And the shadow of the elm must mean the further end of the shadow, otherwise the trunk would have been chosen as the guide. I had then to find where the far end of the shadow would fall when the sun was just clear of the oak.'

'That must have been difficult, Holmes, when the elm was no longer there.'

'Well, at least, I knew that if Brunton could do it, I could also. Besides, there was no real difficulty. I went with Musgrave to his study and whittled myself this peg, to which I tied this long string, with a knot at each yard. Then I took two lengths of a fishing-rod, which came to just six feet, and I went back with my client to where the elm had been. The sun was just grazing the top of the oak. I fastened the rod on end, marked out the direction of the shadow, and measured it. It was 9 ft. in length.

'Of course, the calculation was now a simple one. If a rod of 6 ft. threw a shadow of 9 ft., a tree of 64 ft. would throw one of 96 ft., and the line of one would of course be the line of the other. I measured out the distance, which brought me almost to the wall of the house, and I thrust a peg into the spot. You can imagine my exultation, Watson, when within 2 in. of my peg I saw a conical depression in the ground. I knew that it was the mark made by Brunton in his measurements, and that I was still upon his trail.

'From this starting point I proceeded to step, having first taken the cardinal points by my pocket compass. Ten steps with each foot took me along parallel with the wall of the house, and again I marked my spot with a peg. Then I carefully paced off five to the east and two to the south. It brought me to the very threshold of the old door. Two steps to the west meant now that I was to go two paces down the stone-flagged passage, and this was the place indicated by the Ritual.

'Never have I felt such a cold chill of disappointment, Watson. For a moment it seemed to me that there must be some radical mistake in my calculations. The setting sun shone full upon the passage floor, and I could see that the old foot-worn grey stones, with which it was paved, were firmly cemented together, and had certainly not been moved for many a long year. Brunton had not been at work here. I tapped upon the floor, but it sounded the same all over, and there was no sign of any crack or crevice. But fortunately, Musgrave, who had begun to appreciate the meaning of my proceedings, and who was now as excited as myself, took out his manuscript to check my calculations.

' "And under," he cried: "you have omitted the 'and under'."

'I had thought that it meant that we were to dig, but now, of course, I saw at once that I was wrong. "There is a cellar under this, then?" I cried.

' "Yes, and as old as the house. Down here, through this door."

'We went down a winding stone stair, and my companion, striking a match, lit a large lantern which stood on a barrel in the corner. In an instant it was obvious that we had at last come upon the true place, and that we had not been the only people to visit the spot recently.

'It had been used for the storage of wood, but the billets, which had evidently been littered over the floor, were now piled at the sides so as to leave a clear space in the middle. In this space lay a large and heavy flagstone, with a rusted iron ring in the centre, to which a thick shepherd's check muffler was attached.

' "By Jove!" cried my client, "that's Brunton's muffler. I have seen it on him, and could swear to it. What has the villain been doing here?"

'At my suggestion a couple of the county police were summoned to be present, and I then endeavoured to raise the stone by pulling on the cravat. I could only move it slightly, and it was with the aid of one of the constables that I succeeded at last in carrying it to one side. A black hole yawned beneath, into which we all peered, while Musgrave, kneeling at the side, pushed down the lantern.

'A small chamber about 7 ft. deep and 4 ft. square lay open to us. At one side of this was a squat, brass-bound, wooden box, the lid of which was hinged upwards, with this curious, old-fashioned key projecting from the lock. It was furred outside by a thick layer of dust, and damp and worms had eaten through the wood so that a crop of livid fungi was growing on the inside of it. Several discs of metal—old coins apparently—such as I hold here, were scattered over the bottom of the box, but it contained nothing else.

'At that moment, however, we had no thought for the old chest, for our eyes were riveted upon that which crouched beside it. It was the figure of a man, clad in a suit of black, who squatted down upon his hams with his forehead sunk upon the edge of the box and his two arms thrown out on each side of it. The attitude had drawn all the stagnant blood to his face, and no man could have recognised that distorted, liver-coloured countenance; but his height, his dress, and his hair were all sufficient to show my client, when we had drawn the body up, that it was indeed his missing butler. He had been dead some days, but there was no wound or bruise upon his person to show how he had met his dreadful end. When his body had been carried from the cellar we found ourselves still confronted with a problem which was almost as formidable as that with which we had started.

'I confess that so far, Watson, I had been disappointed in my investigation. I had reckoned upon solving the matter when once I had found the place referred to in the Ritual; but now I was there, and was apparently as far as ever from knowing what it was which the family had concealed with such elaborate precautions. It is true that I had thrown a light upon the fate of Brunton, but now I had to ascertain how that fate had come upon him, and what part had been played in the matter by the woman who had disappeared. I sat down upon a keg in the corner and thought the whole matter carefully over.

'You know my methods in such cases, Watson: I put myself in the man's place, and having first gauged his intelligence, I try to imagine how I should myself have proceeded under the same circumstances. In this case the matter was simplified by Brunton's intelligence being quite first rate, so that it was unnecessary to make any allowance for the personal equation, as the astronomers have dubbed it. He knew that something valuable was concealed. He had spotted the place. He found that the stone which covered it was just too heavy for a man to move unaided. What would he do next? He could not get help from outside, even if he had someone whom he could trust, without the unbarring of doors, and considerable risk of detection. It was better, if he could, to have his helpmate inside the house. But whom could he ask? This girl had been devoted to him. A man always finds it hard to realize that he may have finally lost a woman's love, however badly he may have treated her. He would try by a few attentions to make his peace with the girl Howells, and then would engage her as his accomplice. Together they would come at night to the cellar, and their united force would suffice to raise the stone. So far I could follow their actions as if I had actually seen them.

'But for two of them, and one a woman, it must have been heavy work, the raising of that stone. A burly Sussex policeman and I had found it no light job. What would they do to assist them? Probably what I should have done myself. I rose and examined carefully the different billets of wood which were scattered round the floor. Almost at once I came upon what I expected. One piece, about 3 ft. in length, had a marked indentation at one end, while several were flattened at the sides as if they had been compressed by some considerable weight. Evidently, as they had dragged the stone up they had thrust the chunks of wood into the chink, until at last, when the opening was large enough to crawl through, they would hold it open by a billet placed lengthwise, which might very well become indented at the lower end, since the whole weight of the stone would press it down on to the edge of the other slab. So far I was still on safe ground.

'And now, how was I to proceed to reconstruct this midnight drama? Clearly only one could get into the hole, and that one was Brunton. The girl must have waited above. Brunton then unlocked the box, handed up the contents, presumably—since they were not to be found—and then— and then what happened?

'What smouldering fire of vengeance had suddenly sprung into flame in this passionate Celtic woman's soul when she saw the man who had wronged her—wronged her, perhaps, far more than we suspected—in her power? Was it a chance that the wood had slipped and that the stone had shut Brunton into what had become his sepulchre? Had she only been guilty of silence as to his fate? Or had some sudden blow from her hand dashed the support away and sent the slab crashing down into its place. Be that as it might, I seemed to see that woman's figure, still clutching at her treasure-trove, and flying wildly up the winding stair with her ears ringing perhaps with the muffled screams from behind her, and with the drumming of frenzied hands against the slab of stone which was choking her faithless lover's life out.

'Here was the secret of her blanched face, her shaken nerves, her peals of hysterical laughter on the next morning. But what had been in the box? What had she done with that? Of course, it must have been the old metal and pebbles which my client had dragged from the mere. She had thrown them in there at the first opportunity, to remove the last trace of her crime.

'For twenty minutes I had sat motionless thinking the matter out. Musgrave still stood with a very pale face swinging his lantern and peering down into the hole.

' "These are coins of Charles I,"* said he, holding out the few which had been left in the box. "You see we were right in fixing our date for the Ritual."

' "We may find something else of Charles I," I cried, as the probable meaning of the first two questions of the Ritual broke suddenly upon me. "Let me see the contents of the bag you fished from the mere."

'We ascended to his study, and he laid the *débris* before me. I could understand his regarding it as of small import-ance when I looked at it, for the metal was almost black, and the stones lustreless and dull. I rubbed one of them on my sleeve, however, and it glowed afterwards like a spark, in the dark hollow of my hand. The metal-work was in the form of a double-ring, but it had been bent and twisted out of its original shape.

' "You must bear in mind," said I, "that the Royal party made head in England even after the death of the King, and that when they at last fled they probably left many of their most precious possessions buried behind them, with the intention of returning for them in more peaceful times."

' "My ancestor, Sir Ralph Musgrave, was a prominent Cavalier,* and the right-hand man of Charles II in his wanderings," said my friend.

' "Ah, indeed!" I answered. "Well, now, I think that really should give us the last link that we wanted. I must congratulate you on coming into possession, though in rather a tragic manner, of a relic which is of great intrinsic value, but even of greater importance as an historical curiosity."

' "What is it, then?" he gasped in astonishment.

' "It is nothing less than the ancient crown of the Kings of England."*

' "The crown!"

' "Precisely. Consider what the Ritual says. How does it run? 'Whose was it?' 'His who is gone.' That was after the execution of Charles. Then, 'Who shall have it?' 'He who will come.' That was Charles II, whose advent was already foreseen.* There can, I think, be no doubt that this battered and shapeless diadem once encircled the brows of the Royal Stuarts."

' "And how came it in the pond?"

' "Ah, that is a question which will take some time to answer," and with that I sketched out the whole long chain of surmise and of proof which I had constructed. The twilight had closed in and the moon was shining brightly in the sky before my narrative was finished.

' "And how was it, then, that Charles did not get his crown when he returned?" asked Musgrave, pushing back the relic into its linen bag.

' "Ah, there you lay your finger upon the one point which we shall probably never be able to clear up. It is likely that the Musgrave who held the secret died in the interval, and by some oversight left this guide to his descendant without explaining the meaning of it. From that day to this it has been handed down from father to son, until at last it came within reach of a man who tore its secret out of it and lost his life in the venture."

'And that's the story of the Musgrave Ritual, Watson. They have the crown down at Hurlstone—though they had some legal bother, and a considerable sum to pay* before they were allowed to retain it. I am sure that if you mentioned my name they would be happy to show it to you. Of the woman nothing was ever heard, and the probability is that she got away out of England, and carried herself, and the memory of her crime, to some land beyond the seas.'

The Reigate Squire

IT was some time before the health of my friend, Mr Sherlock Holmes, recovered from the strain caused by his immense exertions in the spring of '87. The whole question of the Netherland-Sumatra Company and of the colossal schemes of Baron Maupertuis* is too recent in the minds of the public, and too intimately concerned with politics and finance, to be a fitting subject for this series of sketches. It led, however, in an indirect fashion to a singular and complex problem, which gave my friend an opportunity of demonstrating the value of a fresh weapon among the many with which he waged his life-long battle against crime.

On referring to my notes, I see that it was on the 14th of April that I received a telegram from Lyons,* which informed me that Holmes was lying ill in the Hotel Dulong. Within twenty-four hours I was in his sick-room, and was relieved to find that there was nothing formidable in his symptoms. His iron constitution, however, had broken down under the strain of an investigation which had extended over two months, during which period he had never worked less than fifteen hours a day, and had more than once, as he assured, me, kept to his task for five days at a stretch. The triumphant issue of his labours could not save him from reaction after so terrible an exertion, and at a time when Europe was ringing with his name and when his room was literally ankle-deep with congratulatory telegrams, I found him a prey to the blackest depression. Even the knowledge that he had succeeded where the police of three countries had failed, and that he had out-manoeuvred at every point the most accomplished swindler in Europe, was insufficient to rouse him from his nervous prostration.

Three days later we were back in Baker Street together, but it was evident that my friend would be much the better for a change, and the thought of a week of spring-time in the country was full of attractions to me also. My old friend, Colonel Hayter, who had come under my professional care in Afghanistan, had now taken a house near Reigate,* in Surrey, and had frequently asked me to come down to him upon a visit. On the last occasion he had remarked that if my friend would only come with me, he would be glad to extend his hospitality to him also. A little diplomacy was needed, but when Holmes understood that the establishment was a bachelor one, and that he would be allowed the fullest freedom, he fell in with my plans, and a week after our return from Lyons we were under the Colonel's roof. Hayter was a fine old soldier, who had seen much of the world, and he soon found, as I had expected, that Holmes and he had plenty in common.

On the evening of our arrival we were sitting in the Colonel's gun-room after dinner, Holmes stretched upon the sofa, while Hayter and I looked over his little armoury of fire-arms.

'By the way,' said he, suddenly, 'I'll take one of these pistols upstairs with me in case we have an alarm.'

'An alarm!' said I.

'Yes, we've had a scare in this part lately. Old Acton, who is one of our county magnates, had his house broken into last Monday. No great damage done, but the fellows are still at large.'

'No clue?' asked Holmes, cocking his eye at the Colonel.

'None as yet. But the affair is a petty one, one of our little country crimes, which must seem too small for your attention, Mr Holmes, after this great international affair.'

Holmes waved away the compliment, though his smile showed that it had pleased him.

'Was there any feature of interest?'

'I fancy not. The thieves ransacked the library, and got very little for their pains. The whole place was turned upside down, drawers burst open and presses ransacked, with the result that an odd volume of Pope's "Homer",* two plated candlesticks, an ivory letter-weight, a small oak barometer, and a ball of twine are all that have vanished.'

'What an extraordinary assortment!' I exclaimed.

'Oh, the fellows evidently grabbed hold of anything they could get.'

Holmes grunted from the sofa.

'The county police ought to make something of that,' said he. 'Why, it is surely obvious that—'

But I held up a warning finger.

'You are here for a rest, my dear fellow. For Heaven's sake, don't get started on a new problem when your nerves are all in shreds.'

Holmes shrugged his shoulders with a glance of comic resignation towards the Colonel, and the talk drifted away into less dangerous channels.

It was destined, however, that all my professional caution should be wasted, for next morning the problem obtruded itself upon us in such a way that it was impossible to ignore it, and our country visit took a turn which neither of us could have anticipated. We were at breakfast when the Colonel's butler rushed in with all his propriety shaken out of him.

'Have you heard the news, sir?' he gasped. 'At the Cunninghams', sir!'

'Burglary?' cried the Colonel, with his coffee cup in mid air.

'Murder!'

The Colonel whistled. 'By Jove!' said he, 'who's killed, then? The J.P.,* or his son?'

'Neither, sir. It was William, the coachman. Shot through the heart, sir, and never spoke again.'

'Who shot him, then?'

'The burglar, sir. He was off like a shot and got clean away. He'd just broke in at the pantry window when William came on him and met his end in saving his master's property.'

'What time?'

'It was last night, sir, somewhere about twelve.'

'Ah, then, we'll step over presently,' said the Colonel, coolly settling down to his breakfast again. 'It's a baddish business,' he added, when the butler had gone. 'He's our leading squire about here, is old Cunningham, and a very decent fellow, too. He'll be cut up over this, for the man has been in his service for years, and was a good servant. It's evidently the same villains who broke into Acton's.'

'And stole that very singular collection?' said Holmes, thoughtfully.

'Precisely.'

'Hum! It may prove the simplest matter in the world; but, all the same, at first glance this is just a little curious, is it not? A gang of burglars acting in the country might be expected to vary the scene of their operations, and not to crack two cribs* in the same district within a few days. When you spoke last night of taking precautions, I remember that it passed through my mind that this was probably the last parish in England to which the thief or thieves would be likely to turn their attention; which shows that I have still much to learn.'

'I fancy it's some local practitioner,' said the Colonel. 'In that case, of course, Acton's and Cunningham's are just the places he would go for, since they are far the largest about here.'

'And richest?'

'Well, they ought to be; but they've had a law-suit for some years which has sucked the blood out of both of them, I fancy. Old Acton has some claim on half Cunningham's estate, and the lawyers have been at it with both hands.'

'If it's a local villain, there should not be much difficulty in running him down,' said Holmes, with a yawn. 'All right, Watson, I don't intend to meddle.'

'Inspector Forrester, sir,' said the butler, throwing open the door.

The official, a smart, keen-faced young fellow, stepped into the room. 'Good morning, Colonel,' said he. 'I hope I don't intrude, but we hear that Mr Holmes, of Baker Street, is here.'

The Colonel waved his hand towards my friend, and the Inspector bowed.

'We thought that perhaps you would care to step across, Mr Holmes.'

'The Fates are against you, Watson,' said he, laughing. 'We were chatting about the matter when you came in, Inspector. Perhaps you can let us have a few details.' As he leaned back in his chair in the familiar attitude, I knew that the case was hopeless.

'We had no clue in the Acton affair. But here we have plenty to go on, and there's no doubt it is the same party in each case. The man was seen.'

'Ah!'

'Yes, sir. But he was off like a deer after the shot that killed poor William Kirwan was fired. Mr Cunningham saw him from the bedroom window, and Mr Alec Cunningham saw him from the back passage. It was a quarter to twelve when the alarm broke out. Mr Cunningham had just got into bed, and Mister Alec* was smoking a pipe in his dressing-gown. They both heard William, the coachman, calling for help, and Mister Alec he ran down to see what was the matter. The back door was open, and as he came to the foot of the stairs he saw two men wrestling together outside. One of them fired a shot, the other dropped, and the murderer rushed across the garden and over the hedge. Mr Cunningham, looking out of his bedroom window, saw the fellow as he gained the road, but lost sight of him at once. Mister Alec stopped to see if he could help the dying man, and so the villain got clean away. Beyond the fact that he was a middle-sized man, and dressed in some dark stuff, we have no personal clue, but we are making energetic inquiries, and if he is a stranger we shall soon find him out.'

'What was this William doing there? Did he say anything before he died?'

'Not a word. He lives at the lodge with his mother, and as he was a very faithful fellow, we imagine that he walked up to the house with the intention of seeing that all was right there. Of course, this Acton business has put everyone on their guard. The robber must have just bust open the door—the lock has been forced—when William came upon him.'

'Did William say anything to his mother before going out?'

'She is very old and deaf, and we can get no information from her. The shock has made her half-witted, but I understand that she was never very bright. There is one very im- portant circumstance, however. Look at this!'

He took a small piece of torn paper from a note-book and spread it out upon his knee.

'This was found between the finger and thumb of the dead man. It appears to be a fragment torn from a larger sheet. You will observe that the hour mentioned upon it is the very time at which the poor fellow met his fate. You see that his murderer might have torn the rest of the sheet from him or he might have taken this fragment from the murderer. It reads almost as though it was an appointment.'

Holmes took up the scrap of paper, a facsimile of which is here reproduced.

'Presuming that it is an appointment,' continued the Inspector, 'it is, of course, a conceivable theory that this William Kirwan, although he had the reputation of being an honest man, may have bee in league with the thief. He may have met him there, may even have helped him to break in the door, and then they may have fallen out between themselves.'

'This writing is of extraordinary interest,' said Holmes, who had been examining it with intense concentration. 'These are much deeper waters than I had thought.' He sank his head upon his hands, while the Inspector smiled at the effect which his case had had upon the famous London specialist.

'Your last remark,' said Holmes, presently, 'as to the possibility of there being an understanding between the burglar and the servant, and this being a note of appointment from one to the other, is an ingenious and not entirely an impossible supposition. But this writing opens up—' he sank his head into his hands again and remained for some minutes in the deepest thought. When he raised his face I was surprised to see that his cheek was tinged with colour, and his eyes as bright as before his illness. He sprang to his feet with all his old energy.

'I'll tell you what!' said he. 'I should like to have a quiet little glance into the details of this case. There is something in it which fascinates me extremely. If you will permit me, Colonel, I will leave my friend, Watson, and you, and I will step round with the Inspector to test the truth of one or two little fancies of mine. I will be with you again in half an hour.'

An hour and a half had elapsed before the Inspector returned alone.

'Mr Holmes is walking up and down in the field outside,' said he. 'He wants us all four to go up to the house together.'

'To Mr Cunningham's?'

'Yes, sir.'

'What for?'

The Inspector shrugged his shoulders. 'I don't quite know, sir. Between ourselves, I think Mr Holmes has not quite got over his illness yet. He's been behaving very queerly, and he is very much excited.'

'I don't think you need alarm yourself,' said I. 'I have usually found that there was method in his madness.'*

'Some folk might say there was madness in his method,' muttered the Inspector. 'But he's all on fire to start, Colonel, so we had best go out, if you are ready.'

We found Holmes pacing up and down in the field, his chin sunk upon his breast, and his hands thrust into his trouser pockets.

'The matter grows in interest,' said he. 'Watson, your country trip has been a distinct success. I have had a charming morning.'

'You have been up to the scene of the crime, I understand?' said the Colonel.

'Yes; the Inspector and I have made quite a little reconnaissance together.'

'Any success?'

'Well, we have seen some very interesting things. I'll tell you what we did as we walk. First of all we saw the body of this unfortunate man. He certainly died from a revolver wound, as reported.'

'Had you doubted it, then?'

'Oh, it is as well to test everything. Our inspection was not wasted. We then had an interview with Mr Cunningham and his son, who were able to point out the exact spot where the murderer had broken through the garden hedge in his flight. That was of great interest.'

'Naturally.'

'Then we had a look at this poor fellow's mother. We could get no information from her, however, as she is very old and feeble.'

'And what is the result of your investigations?'

'The conviction that the crime is a very peculiar one. Perhaps our visit now may do something to make it less obscure. I think that we are both agreed, Inspector, that the fragment of paper in the dead man's hand, bearing, as it does, the very hour of his death written upon it, is of extreme importance.'

'It should give a clue, Mr Holmes.'

'It *does* give a clue. Whoever wrote that note was the man who brought William Kirwan out of his bed at that hour. But where is the rest of that sheet of paper?'

'I examined the ground carefully in the hope of finding it,' said the Inspector.

'It was torn out of the dead man's hand. Why was someone so anxious to get possession of it? Because it incriminated him. And what would he do with it? Thrust it into his pocket most likely, never noticing that a corner of it had been left in the grip of the corpse. If we could get the rest of that sheet, it is obvious that we should have gone a long way towards solving the mystery.'

'Yes, but how can we get at the criminal's pocket before we catch the criminal?'

'Well, well, it was worth thinking over. Then there is another obvious point. The note was sent to William. The man who wrote it could not have taken it, otherwise of course he might have delivered his own message by word of mouth. Who brought the note, then? Or did it come through the post?'

'I have made inquiries,' said the Inspector. 'William received a letter by the afternoon post yesterday. The envelope was destroyed by him.'

'Excellent!' cried Holmes, clapping the Inspector on the back. 'You've seen the postman. It is a pleasure to work with you. Well, here is the lodge, and if you will come up, Colonel, I will show you the scene of the crime.'

We passed the pretty cottage where the murdered man had lived, and walked up an oak-lined avenue to the fine old Queen Anne house, which bears the date of Malplaquet* upon the lintel of the door. Holmes and the Inspector led us round it until we came to the side gate, which is separated by a stretch of garden from the hedge which lines the road. A constable was standing at the kitchen door.

'Throw the door open, officer,' said Holmes. 'Now it was on those stairs that young Mr Cunningham stood and saw the two men struggling just where we are. Old Mr Cunningham was at that window—the second on the left—and he saw the fellow get away just to the left of that bush. So did the son. They are both sure of it, on account of the bush. Then Mister Alec ran out and knelt beside the wounded man. The ground is very hard, you see, and there are no marks to guide us.'

As he spoke two men came down the garden path, from round the angle of the house. The one was an elderly man, with a strong, deep-lined, heavy-eyed face; the other a dashing young fellow, whose bright, smiling expression and showy dress were in strange contrast with the business which had brought us there.

'Still at it, then?' said he to Holmes. 'I thought you Londoners were never at fault. You don't seem to be so very quick, after all.'

'Ah! you must give us a little time,' said Holmes, good-humouredly.

'You'll want it,' said young Alec Cunningham. 'Why, I don't see that we have any clue at all.'

'There's only one,' answered the Inspector. 'We thought that if we could only find—Good heavens! Mr Holmes, what is the matter?'

My poor friend's face had suddenly assumed the most dreadful expression. His eyes rolled upwards, his features writhed in agony, and with a suppressed groan he dropped on his face upon the ground. Horrified at the suddenness and severity of the attack, we carried him into the kitchen, where he lay back in a large chair and breathed heavily for some minutes. Finally, with a shame-faced apology for his weakness, he rose once more.

'Watson would tell you that I have only just recovered from a severe illness,' he explained. 'I am liable to these sudden nervous attacks.'

'Shall I send you home in my trap?' asked old Cunningham.

'Well, since I am here there is one point on which I should like to feel sure. We can very easily verify it.'

'What is it?'

'Well, it seems to me that it is just possible that the arrival of this poor fellow William was not before but after the entrance of the burglar into the house. You appear to take it for granted that although the door was forced the robber never got in.'

'I fancy that is quite obvious,' said Mr Cunningham, gravely. 'Why, my son Alec had not yet gone to bed, and he would certainly have heard anyone moving about.'

'Where was he sitting?'

'I was sitting smoking in my dressing-room.'

'Which window is that?'

'The last on the left, next my father's.'

'Both your lamps were lit, of course?'

'Undoubtedly.'

'There are some very singular points here,' said Holmes, smiling. 'Is it not extraordinary that a burglar—and a burglar who had had some previous experience—should deliberately break into a house at a time when he could see from the lights that two of the family were still afoot?'

'He must have been a cool hand.'

'Well, of course, if the case were not an odd one we should not have been driven to ask you for an explanation,' said Mister Alec. 'But as to your idea that the man had robbed the house before William tackled him, I think it a most absurd notion. Shouldn't we have found the place disarranged and missed the things which he had taken?'

'It depends on what the things were,' said Holmes. 'You must remember that we are dealing with a burglar who is a very peculiar fellow, and who appears to work on lines of his own. Look, for example, at the queer lot of things which he took from Acton's—what was it?—a ball of string, a letter-weight, and I don't know what other odds and ends!'

'Well, we are quite in your hands, Mr Holmes,' said old Cunningham. 'Anything which you or the Inspector may suggest will most certainly be done.'

'In the first place,' said Holmes, 'I should like you to offer a reward—coming from yourself, for the officials may take a little time before they would agree upon the sum, and these things cannot be done too promptly. I have jotted down the form here, if you would not mind signing it. Fifty pounds was quite enough, I thought.'

'I would willingly give five hundred,' said the J.P., taking the slip of paper and the pencil which Holmes handed to him. 'This is not quite correct, however,' he added, glancing over the document.

'I wrote it rather hurriedly.'

'You see you begin: "Whereas, at about a quarter to one on Tuesday morning, an attempt was made"—and so on. It was at a quarter to twelve, as a matter of fact.'

I was pained at the mistake, for I knew how keenly Holmes would feel any slip of the kind. It was his speciality to be accurate as to fact, but his recent illness had shaken him, and this one little incident was enough to show me that he was still far from being himself. He was obviously embarrassed for an instant, while the Inspector raised his eyebrows and Alec Cunningham burst into a laugh. The old gentleman corrected the mistake, however, and handed the paper back to Holmes.

'Get it printed as soon as possible,' he said. 'I think your idea is an excellent one.'

Holmes put the slip of paper carefully away in his pocket-book.

'And now,' said he, 'it would really be a good thing that we should all go over the house together, and make certain that this rather erratic burglar did not, after all, carry anything away with him.'

Before entering, Holmes made an examination of the door which had been forced. It was evident that a chisel or strong knife had been thrust in, and the lock forced back with it. We could see the marks in the wood where it had been pushed in.

'You don't use bars, then?' he asked.

'We have never found it necessary.'

'You don't keep a dog?'

'Yes; but he is chained on the other side of the house.'

'When do the servants go to bed?'

'About ten.'

'I understand that William was usually in bed also at that hour?'

'Yes.'

'It is singular that on this particular night he should have been up. Now, I should be very glad if you would have the kindness to show us over the house, Mr Cunningham.'

A stone-flagged passage, with the kitchens branching away from it, led by a wooden staircase directly to the first floor of the house. It came out upon the landing opposite to a second more ornamental stair which led up from the front hall. Out of this landing opened the drawing-room and several bedrooms, including those of Mr Cunningham and his son. Holmes walked slowly, taking keen note of the architecture of the house. I could tell from his expression that he was on a hot scent, and yet I could not in the least imagine in what direction his inferences were leading him.

'My good sir,' said Mr Cunningham, with some impatience, 'this is surely very unnecessary. That is my room at the end of the stairs, and my son's is the one beyond it. I leave it to your judgment whether it was possible for the thief to have come up here without disturbing us.'

'You must try round and get on a fresh scent, I fancy,' said the son, with a rather malicious smile.

'Still, I must ask you to humour me a little further. I should like, for example, to see how far the windows of the bedrooms command the front. This, I understand, is your son's room'—he pushed open the door—'and that, I presume, is the dressing-room in which he sat smoking when the alarm was given. Where does the window of that look out to?' He stepped across the bedroom, pushed open the door, and glanced round the other chamber.

'I hope you are satisfied now?' said Mr Cunningham testily.

'Thank you; I think I have seen all that I wished.'

'Then, if it is really necessary, we can go into my room.'

'If it is not too much trouble.'

The J.P. shrugged his shoulders, and led the way into his own chamber, which was a plainly furnished and commonplace room. As we moved across it in the direction of the window, Holmes fell back until he and I were the last of the group. Near the foot of the bed was a small square table, on which stood a dish of oranges and a carafe of water. As we passed it, Holmes, to my unutterable astonishment, leaned over in front of me and deliberately knocked the whole thing over. The glass smashed into a thousand pieces, and the fruit rolled about into every corner of the room.

'You've done it now, Watson,' said he, coolly. 'A pretty mess you've made of the carpet.'

I stooped in some confusion and began to pick up the fruit, understanding that for some reason my companion desired me to take the blame upon myself. The others did the same, and set the table on its legs again.

'Hallo!' cried the Inspector, 'where's he got to?'

Holmes had disappeared.

'Wait here an instant,' said young Alec Cunningham. 'The fellow is off his head, in my opinion. Come with me, father, and see where he has got to!'

They rushed out of the room, leaving the Inspector, the Colonel, and me, staring at each other.

' 'Pon my word, I am inclined to agree with Mister Alec,' said the official. 'It may be the effect of this illness, but it seems to me that—'

His words were cut short by a sudden scream of 'Help! Help! Murder!' With a thrill I recognised the voice as that of my friend. I rushed madly from the room on to the landing. The cries, which had sunk down into a hoarse, inarticulate shouting, came from the room which we had first visited. I dashed in, and on into the dressing-room beyond. The two Cunninghams were bending over the prostrate figure of Sherlock Holmes, the younger clutching his throat with both hands, while the elder seemed to be twisting one of his wrists. In an instant the three of us had torn them away from him, and Holmes staggered to his feet, very pale, and evidently greatly exhausted.

'Arrest these men, Inspector!' he gasped.

'On what charge?'

'That of murdering their coachman, William Kirwan!'

The Inspector stared about him in bewilderment. 'Oh, come now, Mr Holmes,' said he at last; 'I am sure you don't really mean to—'

'Tut, man; look at their faces!' cried Holmes curtly.

Never, certainly, have I seen a plainer confession of guilt upon human countenances. The older man seemed numbed and dazed, with a heavy, sullen expression upon his strongly marked face. The son, on the other hand, had dropped all that jaunty, dashing style which had characterized him, and the ferocity of a dangerous wild beast gleamed in his dark eyes and distorted his handsome features. The Inspector said nothing, but, stepping to the door, he blew his whistle. Two of his constables came at the call.

'I have no alternative, Mr Cunningham,' said he. 'I trust that this may all prove to be an absurd mistake; but you can see that—Ah, would you? Drop it!' He struck out with his hand, and a revolver, which the younger man was in the act of cocking, clattered down upon the floor.

'Keep that,' said Holmes, quickly putting his foot upon it. 'You will find it useful at the trial. But this is what we really wanted.' He held up a little crumpled piece of paper.

'The remainder of the sheet!' cried the Inspector.

'Precisely.'

'And where was it?'

'Where I was sure it must be. I'll make the whole matter clear to you presently. I think, Colonel, that you and Watson might return now, and I will be with you again in an hour at the furthest. The Inspector and I must have a word with the prisoners; but you will certainly see me back at luncheon time.'

Sherlock Holmes was as good as his word, for about one o'clock he rejoined us in the Colonel's smoking-room. He was accompanied by a little, elderly gentleman, who was introduced to me as the Mr Acton whose house had been the scene of the original burglary.

'I wished Mr Acton to be present while I demonstrated this small matter to you,' said Holmes, 'for it is natural that he should take a keen interest in the details. I am afraid, my dear Colonel, that you must regret the hour that you took in such a stormy petrel* as I am.'

'On the contrary,' answered the Colonel, warmly, 'I consider it the greatest privilege to have been permitted to study your methods of working. I confess that they quite surpass my expectations, and that I am utterly unable to account for your result. I have not yet seen the vestige of a clue.'

'I am afraid that my explanation may disillusionize you, but it has always been my habit to hide none of my methods, either from my friend Watson or from anyone who might take an intelligent interest in them. But first, as I am rather shaken by the knocking about which I had in the dressing-room, I think that I shall help myself to a dash of your brandy, Colonel. My strength has been rather tried of late.'

'I trust you had no more of those nervous attacks.'

Sherlock Holmes laughed heartily. 'We will come to that in its turn,' said he. 'I will lay an account of the case before you in its due order, showing you the various points which guided me in my decision. Pray interrupt me if there is any inference which is not perfectly clear to you.

'It is of the highest importance in the art of detection to be able to recognise out of a number of facts which are incidental and which vital. Otherwise your energy and attention must be dissipated instead of being concentrated. Now, in this case there was not the slightest doubt in my mind from the first that the key of the whole matter must be looked for in the scrap of paper in the dead man's hand.

'Before going into this I would draw your attention to the fact that if Alec Cunningham's narrative were correct, and if the assailant after shooting William Kirwan had *instantly* fled, then it obviously could not be he who tore the paper from the dead man's hand. But if it was not he, it must have been Alec Cunningham himself, for by the time the old man had descended several servants were upon the scene. The point is a simple one, but the Inspector had overlooked it because he had started with the supposition that these county magnates had had nothing to do with the matter. Now, I make a point of never having any prejudices and of following docilely wherever fact may lead me, and so in the very first stage of the investigation I found myself looking a little askance at the part which had been played by Mr Alec Cunningham.

'And now I made a very careful examination of the corner of paper which the Inspector had submitted to us. It was at once clear to me that it formed part of a very remarkable document. Here it is. Do you not now observe something very suggestive about it?'

'It has a very irregular look,' said the Colonel.

'My dear sir,' cried Holmes, 'there cannot be the least doubt in the world that it has been written by two persons doing alternate words. When I draw your attention to the strong t's of "at" and "to" and ask you to compare them with the weak ones of "quarter" and "twelve," you will instantly recognise the fact. A very brief analysis of those four words would enable you to say with the utmost confidence that the "learn" and the "maybe" are written in the stronger hand, and the "what" in the weaker.'

'By Jove, it's as clear as day!' cried the Colonel. 'Why on earth should two men write a letter in such a fashion?'

'Obviously the business was a bad one, and one of the men who distrusted the other was determined that, whatever was done, each should have an equal hand in it. Now, of the two men it is clear that the one who wrote the "at" and "to" was the ringleader.'

'How do you get at that?'

'We might deduce it from the mere character of the one hand as compared with the other. But we have more assured reasons than that for supposing it. If you examine this scrap with attention you will come to the conclusion that the man with the stronger hand wrote all his words first, leaving blanks for the other to fill up. These blanks were not always sufficient, and you can see that the second man had a squeeze to fit his "quarter" in between the "at" and the "to", showing that the latter were already written. The man who wrote all his words first is undoubtedly the man who planned this affair.'

'Excellent!' cried Mr Acton.

'But very superficial,' said Holmes. 'We come now, however, to a point which is of importance. You may not be aware that the deduction of a man's age from his writing is one which has been brought to considerable accuracy by experts.* In normal cases one can place a man in his true decade with tolerable confidence. I say normal cases, because ill-health and physical weakness reproduce the signs of old age, even when the invalid is a youth. In this case, looking at the bold, strong hand of the one, and the rather broken-backed appearance of the other, which still retains its legibility, although the t's have begun to lose their crossings, we can say that the one was a young man, and the other was advanced in years without being positively decrepit.'

'Excellent!' cried Mr Acton again.

'There is a further point, however, which is subtler and of greater interest. There is something in common between these hands. They belong to men who are blood-relatives. It may be most obvious to you in the Greek e's, but to me there are many small points which indicate the same thing. I have no doubt at all that a family mannerism can be traced in these two specimens of writing. I am only, of course, giving you the leading results now of my examination of the paper. There were twenty-three other deductions which would be of more interest to experts than to you. They all tended to deepen the impression upon my mind that the Cunninghams, father and son, had written this letter.

'Having got so far, my next step was, of course, to examine into the details of the crime and to see how far they would help us. I went up to the house with the Inspector, and saw all that was to be seen. The wound upon the dead man was, as I was able to determine with absolute confidence, caused by a shot from a revolver fired at the distance of something over four yards. There was no powder-blackening on the clothes. Evidently, therefore, Alec Cunningham had lied when he said that the two men were struggling when the shot was fired. Again, both father and son agreed as to the place where the man escaped into the road. At that point, however, as it happens, there is a broadish ditch, moist at the bottom. As there were no indications of boot-marks about this ditch, I was absolutely sure not only that the Cunninghams had again lied, but that there had never been any unknown man upon the scene at all.

'And now I had to consider the motive of this singular crime. To get at this I endeavoured first of all to solve the reason of the original burglary at Mr Acton's. I understood from something which the Colonel told us that a law-suit had been going on between you, Mr Acton, and the Cunninghams. Of course, it instantly occurred to me that they had broken into your library with the intention of getting at some document which might be of importance in the case.'

'Precisely so,' said Mr Acton; 'there can be no possible doubt as to their intentions. I have the clearest claim upon half their present estate, and if they could have found a single paper*—which, fortunately, was in the strong box of my solicitors—they would undoubtedly have crippled our case.'

'There you are!' said Holmes, smiling. 'It was a dangerous, reckless attempt, in which I seem to trace the influence of young Alec. Having found nothing, they tried to divert suspicion by making it appear to be an ordinary burglary, to which end they carried off whatever they could lay their hands upon. That is all clear enough, but there was much that was still obscure. What I wanted above all was to get the missing part of that note. I was certain that Alec had torn it out of the dead man's hand, and almost certain that he must have thrust it into the pocket of his dressing-gown. Where else could he have put it? The only question was whether it was still there. It was worth an effort to find out, and for that object we all went up to the house.

'The Cunninghams joined us, as you doubtless remember, outside the kitchen door. It was, of course, of the very first importance that they should not be reminded of the existence of this paper, otherwise they would naturally destroy it without delay. The Inspector was about to tell them the importance which we attached to it when, by the luckiest chance in the world, I tumbled down in a sort of fit and so changed the conversation.'

'Good heavens!' cried the Colonel, laughing. 'Do you mean to say all our sympathy was wasted and your fit an imposture?'

'Speaking professionally, it was admirably done,' cried I, looking in amazement at this man who was for ever confounding me with some new phase of his astuteness.

'It is an art which is often useful,' said he. 'When I recovered I managed by a device, which had, perhaps, some little merit of ingenuity, to get old Cunningham to write the word "twelve", so that I might compare it with the "twelve" upon the paper.'

'Oh, what an ass I have been!' I exclaimed.

'I could see that you were commiserating with me over my weakness,' said Holmes, laughing. 'I was sorry to cause you the sympathetic pain which I know that you felt. We then went upstairs together, and having entered the room and seen the dressing-gown hanging up behind the door, I contrived by upsetting a table to engage their attention for the moment and slipped back to examine the pockets. I had hardly got the paper, however, which was as I had expected, in one of them, when the two Cunninghams were on me, and would, I verily believe, have murdered me then and there but for your prompt and friendly aid. As it is, I feel that young man's grip on my throat now, and the father has twisted my wrist round in the effort to get the paper out of my hand. They saw that I must know all about it, you see, and the sudden change from absolute security to complete despair made them perfectly desperate.

'I had a little talk with old Cunningham afterwards as to the motive of the crime. He was tractable enough, though his son was a perfect demon, ready to blow out his own or anybody else's brains if he could have got to his revolver. When Cunningham saw that the case against him was so strong he lost all heart, and made a clean breast of everything. It seems that William had secretly followed his two masters on the night when they made their raid upon Mr Acton's, and, having thus got them into his power, proceeded under threats of exposure to levy blackmail upon them. Mister Alec, however, was a dangerous man to play games of that sort with. It was a stroke of positive genius on his part to see in the burglary scare, which was convulsing the country-side, an opportunity of plausibly getting rid of the man whom he feared. William was decoyed up and shot; and, had they only got the whole of the note, and paid a little more attention to detail in their accessories, it is very possible that suspicion might never have been aroused.'

'And the note?' I asked.

Sherlock Holmes placed the subjoined paper before us:

> If you will only come round (at quarter to twelve to the east gate you will learn what will very much surprise you and maybe be of the greatest service to you and also to Annie Morrison. But say nothing to anyone upon the matter

'It is very much the sort of thing that I expected,' said he. 'Of course, we do not yet know what the relations may have been between Alec Cunningham, William Kirwan, and Annie Morrison. The result shows that the trap was skilfully baited. I am sure that you cannot fail to be delighted with the traces of heredity shown in the p's and in the tails of the g's. The absence of the i-dots in the old man's writing is also most characteristic. Watson, I think our quiet rest in the country has been a distinct success, and I shall certainly return, much invigorated, to Baker Street to-morrow.'

The Crooked Man

ONE summer night, a few months after my marriage, I was seated by my own hearth smoking a last pipe and nodding over a novel, for my day's work had been an exhausting one. My wife had already gone upstairs, and the sound of the locking of the hall door some time before told me that the servants had also retired. I had risen from my seat and was knocking out the ashes of my pipe, when I suddenly heard the clang of the bell.

I looked at the clock. It was a quarter to twelve. This could not be a visitor at so late an hour. A patient, evidently, and possibly an all-night sitting. With a wry face I went out into the hall and opened the door. To my astonishment, it was Sherlock Holmes who stood upon my step.

'Ah, Watson,' said he, 'I hoped that I might not be too late to catch you.'

'My dear fellow, pray come in.'

'You look surprised, and no wonder! Relieved, too, I fancy! Hum! you still smoke the Arcadia mixture of your bachelor days, then! There's no mistaking that fluffy ash upon your coat. It's easy to tell that you've been accustomed to wear a uniform, Watson; you'll never pass as a pure-bred civilian as long as you keep that habit of carrying your handkerchief in your sleeve.* Could you put me up to-night?'

'With pleasure.'

'You told me that you had bachelor quarters for one, and I see that you have no gentleman visitor at present. Your hat-stand proclaims as much.'

'I shall be delighted if you will stay.'

'Thank you. I'll fill a vacant peg, then. Sorry to see that you've had the British workman* in the house. He's a token of evil. Not the drains, I hope?'

'No, the gas.'

'Ah! He has left two nail marks from his boot upon your linoleum just where the light strikes it. No, thank you, I had some supper at Waterloo, but I'll smoke a pipe with you with pleasure.'

I handed him my pouch, and he seated himself opposite to me, and smoked for some time in silence. I was well aware that nothing but business of importance could have brought him to me at such an hour, so I waited patiently until he should come round to it.

'I see that you are professionally rather busy just now,' said he, glancing very keenly across at me.

'Yes, I've had a busy day,' I answered. 'It may seem very foolish in your eyes,' I added, 'but really I don't know how you deduced it.'

Holmes chuckled to himself.

'I have the advantage of knowing your habits, my dear Watson,' said he. 'When your round is a short one you walk, and when it is a long one you use a hansom.* As I perceive that your boots, although used, are by no means dirty, I cannot doubt that you are at present busy enough to justify the hansom.'

'Excellent!' I cried.

'Elementary,'* said he. 'It is one of those instances where the reasoner can produce an effect which seems remarkable to his neighbour, because the latter has missed the one little point which is the basis of the deduction. The same may be said, my dear fellow, for the effect of some of these little sketches of yours, which is entirely meretricious, depending as it does upon your retaining in your own hands some factors in the problem which are never imparted to the reader. Now, at present I am in the position of these same readers, for I hold in this hand several threads of one of the strangest cases which ever perplexed a man's brain, and yet I lack the one or two which are needful to complete my theory. But I'll have them, Watson, I'll have them!' His eyes kindled and a slight flush sprang into his thin cheeks. For an instant the veil had lifted upon his keen, intense nature, but for an instant only. When I glanced again his face had resumed that Red Indian composure which had made so many regard him as a machine rather than a man.

'The problem presents features of interest,' said he; 'I may even say very exceptional features of interest. I have already looked into the matter, and have come, as I think, within sight of my solution. If you could accompany me in that last step, you might be of considerable service to me.'

'I should be delighted.'

'Could you go as far as Aldershot to-morrow?'

'I have no doubt Jackson would take my practice.'

'Very good. I want to start by the 11.10 from Waterloo.'

'That would give me time.'

'Then, if you are not too sleepy, I will give you a sketch of what has happened and of what remains to be done.'

'I was sleepy before you came. I am quite wakeful now.'

'I will compress the story as far as may be done without omitting anything vital to the case. It is conceivable that you may even have read some account of the matter. It is the supposed murder of Colonel Barclay, of the Royal Mallows,* at Aldershot, which I am investigating.'

'I have heard nothing of it.'

'It has not excited much attention yet, except locally. The facts are only two days old. Briefly they are these:

'The Royal Mallows is, as you know, one of the most famous Irish regiments in the British Army. It did wonders both in the Crimea and the Mutiny,* and has since that time distinguished itself upon every possible occasion. It was commanded up to Monday night by James Barclay, a gallant veteran, who started as a full private, was raised to commissioned rank for his bravery at the time of the Mutiny, and so lived to command the regiment in which he had once carried a musket.

'Colonel Barclay had married at the time when he was a sergeant, and his wife, whose maiden name was Miss Nancy Devoy,* was the daughter of a former colour-sergeant* in the same corps. There was, therefore, as can be imagined, some little social friction when the young couple (for they were still young) found themselves in their new surroundings.

They appear, however, to have quickly adapted themselves, and Mrs Barclay has always, I understand, been as popular with the ladies of the regiment as her husband was with his brother officers. I may add that she was a woman of great beauty, and that even now, when she has been married for upwards of thirty years, she is still of a striking appearance.

'Colonel Barclay's family life appears to have been a uniformly happy one. Major Murphy, to whom I owe most of my facts, assures me that he has never heard of any misunderstanding between the pair. On the whole he thinks that Barclay's devotion to his wife was greater than his wife's to Barclay. He was acutely uneasy if he were absent from her for a day. She, on the other hand, though devoted and faithful, was less obtrusively affectionate. But they were regarded in the regiment as the very model of a middle-aged couple. There was absolutely nothing in their mutual relations to prepare people for the tragedy which was to follow.

'Colonel Barclay himself seems to have had some singular traits in his character. He was a dashing, jovial old soldier in his usual mood, but there were occasions on which he seemed to show himself capable of considerable violence and vindictiveness. This side of his nature, however, appears never to have been turned towards his wife. Another fact which had struck Major Murphy, and three out of five of the other officers with whom I conversed, was the singular sort of depression which came upon him at times. As the Major expressed it, the smile had often been struck from his mouth, as if by some invisible hand, when he has been joining in the gaieties and chaff of the mess table. For days on end, when the mood was on him, he had been sunk* in the deepest gloom. This and a certain tinge of superstition were the only unusual traits in his character which his brother officers had observed. The latter peculiarity took the form of a dislike to being left alone, especially after dark. This puerile feature in a nature which was conspicuously manly had often given rise to comment and conjecture.

'The first battalion of the Royal Mallows (which is the old 117th) has been stationed at Aldershot for some years. The married officers live out of barracks, and the Colonel has during all this time occupied a villa called Lachine, about half a mile from the North Camp. The house stands in its own grounds, but the west side of it is not more than thirty yards from the high road. A coachman and two maids form the staff of servants. These, with their master and mistress, were the sole occupants of Lachine, for the Barclays had no children, nor was it usual for them to have resident visitors.

'Now for the events at Lachine between nine and ten on the evening of last Monday.

'Mrs Barclay was, it appears, a member of the Roman Catholic Church, and had interested herself very much in the establishment of the Guild of St George,* which was formed in connection with the Watt Street Chapel for the purpose of supplying the poor with cast-off clothing. A meeting of the Guild had been held that evening at eight, and Mrs Barclay had hurried over her dinner in order to be present at it. When leaving the house, she was heard by the coachman to make some commonplace remark to her husband, and to assure him that she would be back before long. She then called for Miss Morrison, a young lady who lives in the next villa, and the two went off together to their meeting. It lasted forty minutes, and at a quarter-past nine Mrs Barclay returned home, having left Miss Morrison at her door as she passed.

'There is a room which is used as a morning-room at Lachine. This faces the road, and opens by a large glass folding door on to the lawn. The lawn is thirty yards across, and is only divided from the highway by a low wall with an iron rail above it. It was into this room that Mrs Barclay went upon her return. The blinds were not down, for the room was seldom used in the evening, but Mrs Barclay herself lit the lamp and then rang the bell, asking Jane Stewart, the housemaid, to bring her a cup of tea, which was quite contrary to her usual habits. The Colonel had been sitting in the dining-room, but hearing that his wife had returned, he joined her in the morning-room. The coachman saw him cross the hall, and enter it. He was never seen again alive.

'The tea which had been ordered was brought up at the end of ten minutes, but the maid, as she approached the door, was surprised to hear the voices of her master and mistress in furious altercation. She knocked without receiving any answer, and even turned the handle, but only to find that the door was locked upon the inside. Naturally enough, she ran down to tell the cook, and the two women with the coachman came up into the hall and listened to the dispute which was still raging. They all agree that only two voices were to be heard, those of Barclay and his wife.* Barclay's remarks were subdued and abrupt, so that none of them were audible to the listeners. The lady's, on the other hand, were most bitter, and, when she raised her voice, could be plainly heard. "You coward!" she repeated over and over again. "What can be done now? Give me back my life. I will never so much as breathe the same air as you again! You coward! You coward!" Those were scraps of her conversation, ending in a sudden dreadful cry in the man's voice, with a crash, and a piercing scream from the woman. Convinced that some tragedy had occurred, the coachman rushed to the door and strove to force it, while scream after scream issued from within. He was unable, however, to make his way in, and the maids were too distracted with fear to be of any assistance to him. A sudden thought struck him, however, and he ran through the hall door and round to the lawn, upon which the long french windows opened. One side of the window was open, which I understand was quite usual in the summer-time, and he passed without difficulty into the room. His mistress had ceased to scream, and was stretched insensible upon a couch, while with his feet tilted over the side of an armchair, and his head upon the ground near the corner of the fender, was lying the unfortunate soldier, stone dead, in a pool of his own blood.

'Naturally the coachman's first thought, on finding that he could do nothing for his master, was to open the door. But here an unexpected and singular difficulty presented itself. The key was not on the inner side of the door, nor could he find it anywhere in the room. He went out again, therefore, through the window, and having obtained the help of a policeman and of a medical man, he returned. The lady, against whom naturally the strongest suspicion rested, was removed to her room, still in a state of insensibility. The Colonel's body was then placed upon the sofa, and a careful examination made of the scene of the tragedy.

'The injury from which the unfortunate veteran was suffering was found to be a ragged cut, some two inches long, at the back part of his head, which had evidently been caused by a violent blow from a blunt weapon. Nor was it difficult to guess what that weapon may have been. Upon the floor, close to the body, was lying a singular club of hard carved wood with a bone handle. The Colonel possessed a varied collection of weapons brought from the different countries in which he had fought, and it is conjectured by the police that this club was among his trophies. The servants deny having seen it before, but among the numerous curiosities in the house it is possible that it may have been overlooked. Nothing else of importance was discovered in the room by the police, save the inexplicable fact that neither upon Mrs Barclay's person, nor upon that of the victim, nor in any part of the room was the missing key to be found. The door had eventually to be opened by a locksmith from Aldershot.

'That was the state of things, Watson, when upon the Tuesday morning I, at the request of Major Murphy, went down to Aldershot to supplement the efforts of the police. I think you will acknowledge that the problem was already one of interest, but my observations soon made me realize that it was in truth much more extraordinary than would at first sight appear.

'Before examining the room I cross-questioned the servants, but only succeeded in eliciting the facts which I have already stated. One other detail of interest was remembered by Jane Stewart, the housemaid. You will remember that on hearing the sound of the quarrel she descended and returned with the other servants. On that first occasion, when she was alone, she says that the voices of her master and mistress were sunk so low that she could hear hardly anything, and judged by their tones, rather than their words, that they had fallen out. On my pressing her, however, she remembered that she heard the word "David" uttered twice by the lady. The point is of the utmost importance as guiding us towards the reason of the sudden quarrel. The Colonel's name, you remember, was James.

'There was one thing in the case which had made the deepest impression both upon the servants and the police. This was the contortion of the Colonel's face. It had set, according to their account, into the most dreadful expression of fear and horror which a human countenance is capable of assuming. More than one person fainted at the mere sight of him, so terrible was the effect. It was quite certain that he had foreseen his fate, and that it had caused him the utmost horror. This, of course, fitted in well enough with the police theory, if the Colonel could have seen his wife making a murderous attack upon him. Nor was the fact of the wound being on the back of his head a fatal objection to this, as he might have turned to avoid the blow. No information could be got from the lady herself, who was temporarily insane from an acute attack of brain fever.

'From the police I learned that Miss Morrison, who, you remember, went out that evening with Mrs Barclay, denied having any knowledge of what it was which had caused the ill-humour in which her companion had returned.

'Having gathered these facts, Watson, I smoked several pipes over them, trying to separate those which were crucial from others which were merely incidental. There could be no question that the most distinctive and suggestive point in the case was the singular disappearance of the door key. A most careful search had failed to discover it in the room. Therefore, it must have been taken from it. But neither the Colonel nor the Colonel's wife could have taken it. That was perfectly clear. Therefore a third person must have entered the room. And that third person could only have come in through the window. It seemed to me that a careful examination of the room and the lawn might possibly reveal some traces of this mysterious individual. You know my methods, Watson. There was not one of them which I did not apply to the inquiry. And it ended by my discovering traces, but very different ones from those which I had expected. There had been a man in the room, and he had crossed the lawn coming from the road. I was able to obtain five very clear impressions of his footmarks—one on the roadway itself, at the point where he had climbed the low wall, two on the lawn, and two very faint ones upon the stained boards near the window where he had entered. He had apparently rushed across the lawn, for his toe marks were much deeper that his heels. But it was not the man who surprised me. It was his companion.'

'His companion!'

Holmes pulled a large sheet of tissue paper out of his pocket and carefully unfolded it upon his knee.

'What do you make of that?' he asked.

The paper was covered with tracings of the footmarks of some small animal. It had five well-marked footpads, an indication of long nails, and the whole print might be nearly as large as a dessert spoon.

'It's a dog,' said I.

'Did you ever hear of a dog running up a curtain? I found distinct traces that this creature had done so.'

'A monkey, then?'

'But it is not the print of a monkey.'

'What can it be, then?'

'Neither dog, nor cat, nor monkey, nor any creature that we are familiar with. I have tried to reconstruct it from the measurements. Here are four prints where the beast has been standing motionless. You see that it is no less than fifteen inches from fore foot to hind. Add to that the length of neck and head, and you get a creature not much less than two feet long—probably more if there is any tail. But now observe this other measurement. The animal has been moving, and we have the length of its stride. In each case it is only about three inches. You have an indication, you see, of a long body with very short legs attached to it. It has not been considerate enough to leave any of its hair behind it. But its general shape must be what I have indicated, and it can run up a curtain and is carnivorous.'

'How do you deduce that?'

'Because it ran up the curtain. A canary's cage was hanging in the window, and its aim seems to have been to get at the bird.'

'Then what was the beast?'

'Ah, if I could give it a name it might go a long way towards solving the case. On the whole it was probably some creature of the weasel or stoat tribe—and yet it is larger than any of these that I have seen.'

'But what had it to do with the crime?'

'That also is still obscure. But we have learned a good deal, you perceive. We know that a man stood in the road looking at the quarrel between the Barclays—the blinds were up and the room lighted. We know also that he ran across the lawn, entered the room, accompanied by a strange animal, and that he either struck the Colonel, or, as is equally possible, that the Colonel fell down from sheer fright at the sight of him, and cut his head on the corner of the fender. Finally, we have the curious fact that the intruder carried away the key with him when he left.'

'Your discoveries seem to have left the business more obscure than it was before,' said I.

'Quite so. They undoubtedly showed that the affair was much deeper than was at first conjectured. I thought the matter over, and I came to the conclusion that I must approach the case from another aspect. But really, Watson, I am keeping you up, and I might just as well tell you all this on our way to Aldershot to-morrow.'

'Thank you, you've gone rather too far to stop.'

'It was quite certain that when Mrs Barclay left the house at half-past seven she was on good terms with her husband. She was never, as I think I have said, ostentatiously affectionate, but she was heard by the coachman chatting with the Colonel in a friendly fashion. Now, it was equally certain that immediately on her return she had gone to the room in which she was least likely to see her husband, had flown to tea, as an agitated woman will, and, finally, on his coming in to her, had broken into violent recriminations. Therefore, something had occurred between seven-thirty and nine o'clock which had completely altered her feelings towards him. But Miss Morrison had been with her during the whole of that hour and a half. It was absolutely certain, therefore, in spite of her denial, that she must know something of the matter.

'My first conjecture was that possibly there had been some passages between this young woman and the old soldier, which the former had now confessed to the wife. That would account for the angry return and also for the girl's denial that anything had occurred. Nor would it be entirely incompatible with most of the words overheard. But there was the reference to David, and there was the known affection of the Colonel for his wife to weigh against it, to say nothing of the tragic intrusion of this other man, which might, of course, be entirely disconnected with what had gone before. It was not easy to pick one's steps, but on the whole I was inclined to dismiss the idea that there had been anything between the Colonel and Miss Morrison, but more than ever convinced that the young lady held the clue as to what it was which had turned Mrs Barclay to hatred of her husband. I took the obvious course, therefore, of calling upon Miss Morrison, of explaining to her that I was perfectly certain that she held the facts in her possession, and of assuring her that her friend, Mrs Barclay, might find herself in the dock upon a capital charge unless the matter were cleared up.

'Miss Morrison is a little, ethereal slip of a girl, with timid eyes and blonde hair, but I found her by no means wanting in shrewdness and common sense. She sat thinking for some time after I had spoken, and then turning to me with a brisk air of resolution, she broke into a remarkable statement, which I will condense for your benefit.

' "I promised my friend that I would say nothing of the matter, and a promise is a promise," said she. "But if I can really help her when so serious a charge is made against her, and when her own mouth, poor darling, is closed by illness, then I think I am absolved from my promise. I will tell you exactly what happened on Monday evening.

' "We were returning from the Watt Street Mission, about a quarter to nine o'clock. On our way we had to pass through Hudson Street,* which is a very quiet thoroughfare. There is only one lamp in it upon the left-hand side, and as we approached this lamp I saw a man coming towards us with his back very bent, and something like a box slung over one of his shoulders. He appeared to be deformed, for he carried his head low, and walked with his knees bent. We were passing him when he raised his face to look at us in the circle of light thrown by the lamp, and as he did so he stopped and screamed out in a dreadful voice, 'My God, it's Nancy!' Mrs Barclay turned as white as death, and would have fallen down had the dreadful-looking creature not caught hold of her. I was going to call for the police, but she, to my surprise, spoke quite civilly to the fellow.

' " 'I thought you had been dead this thirty years, Henry,' said she, in a shaking voice.

' " 'So I have,' said he, and it was awful to hear the tones that he said it in. He had a very dark, fearsome face, and a gleam in his eyes that comes back to me in my dreams. His hair and whiskers were shot with grey, and his face was all crinkled and puckered like a withered apple.

' " 'Just walk on a little way, dear,' said Mrs Barclay. 'I want to have a word with this man. There is nothing to be afraid of.' She tried to speak boldly, but she was still deadly pale, and could hardly get her words out for the trembling of her lips.

' "I did as she asked me, and they talked together for a few minutes. Then she came down the street with her eyes blazing, and I saw the crippled wretch standing by the lamp-post and shaking his clenched fists in the air, as if he were mad with rage. She never said a word until we were at the door here, when she took me by the hand and begged me to tell no one what had happened. 'It is an old acquaintance of mine who has come down in the world,' said she. When I promised her that I would say nothing she kissed me, and I have never seen her since. I have told you now the whole truth, and if I withheld it from the police it is because I did not realize then the danger in which my dear friend stood. I know that it can only be to her advantage that everything should be known."

'There was her statement, Watson, and to me, as you can imagine, it was like a light on a dark night. Everything which had been disconnected before began at once to assume its true place, and I had a shadowy presentiment of the whole sequence of events. My next step obviously was to find the man who had produced such a remarkable impression upon Mrs Barclay. If he were still in Aldershot it should not be a very difficult matter. There are not such a very great number of civilians, and a deformed man was sure to have attracted attention. I spent a day in the search, and by evening—this very evening, Watson—I had run him down. The man's name is Henry Wood, and he lives in lodgings in the same street in which the ladies met him. He has only been five days in the place. In the character of a registration agent* I had a most interesting gossip with his landlady. The man is by trade a conjurer and performer, going round the canteens, after nightfall, and giving a little entertainment at each. He carries some creature about with him in his box,* about which the landlady seemed to be in considerable trepidation, for she had never seen an animal like it. He uses it in some of his tricks, according to her account. So much the woman was able to tell me, and also that it was a wonder the man lived, seeing how twisted he was, and that he spoke in a strange tongue sometimes, and that for the last two nights she had heard him groaning and weeping in his bedroom. He was all right as far as money went, but in his deposit he had given her what looked like a bad florin.* She showed it to me, Watson, and it was an Indian rupee.*

'So now, my dear fellow, you see exactly how we stand and why it is I want you. It is perfectly plain that after the ladies parted from this man he followed them at a distance, that he saw the quarrel between husband and wife through the window, that he rushed in, and that the creature which he carried in his box got loose. That is all very certain. But he is the only person in this world who can tell us exactly what happened in that room.'

'And you intend to ask him?'

'Most certainly—but in the presence of a witness.'

'And I am the witness?'

'If you will be so good. If he can clear the matter up, well and good. If he refuses, we have no alternative but to apply for a warrant.'

'But how do you know he will be there when we return?'

'You may be sure that I took some precautions. I have one of my Baker Street boys* mounting guard over him who would stick to him like a burr, go where he might. We shall find him in Hudson Street to-morrow, Watson; and meanwhile I should be the criminal myself if I kept you out of bed any longer.'

It was midday when we found ourselves at the scene of the tragedy, and, under my companion's guidance, we made our way at once to Hudson Street. In spite of his capacity for concealing his emotions I could easily see that Holmes was in a state of suppressed excitement, while I was myself tingling with that half-sporting, half-intellectual pleasure which I invariably experienced when I associated myself with him in his investigations.

'This is the street,' said he, as he turned into a short thoroughfare lined with plain, two-storied brick houses— 'Ah! here is Simpson to report.'

'He's in all right, Mr Holmes,' cried a small street Arab,* running up to us.

'Good, Simpson!' said Holmes, patting him on the head. 'Come along, Watson. This is the house.' He sent in his card with a message that he had come on important business, and a moment later we were face to face with the man whom we had come to see. In spite of the warm weather he was crouching over a fire, and the little room was like an oven. The man sat all twisted and huddled in his chair in a way which gave an indescribable impression of deformity, but the face which he turned towards us, though worn and swarthy, must at some time have been remarkable for its beauty. He looked suspiciously at us now out of yellow-shot bilious eyes, and, without speaking or rising, he waved towards two chairs.

'Mr Henry Wood, late of India, I believe?' said Holmes, affably. 'I've come over this little matter of Colonel Barclay's death.'

'What should I know about that?'

'That's what I wanted to ascertain. You know, I suppose, that unless the matter is cleared up, Mrs Barclay, who is an old friend of yours, will in all probability be tried for murder?'

The man gave a violent start.

'I don't know who you are,' he cried, 'nor how you come to know what you do know; but will you swear that this is true that you tell me?'

'Why, they are only waiting for her to come to her senses to arrest her.'

'My God! Are you in the police yourself?'

'No.'

'What business is it of yours, then?'

'It's every man's business to see justice done.'

'You can take my word that she is innocent.'

'Then you are guilty?'

'No, I am not.'

'Who killed Colonel James Barclay, then?'

'It was a just Providence that killed him. But mind you this, that if I had knocked his brains out, as it was in my heart to do, he would have had no more than his due from my hands. If his own guilty conscience had not struck him down, it is likely enough that I might have had his blood upon my soul. You want me to tell the story? Well, I don't know why I shouldn't, for there's no cause for me to be ashamed of it.

'It was in this way, sir. You see me now with my back like a camel and my ribs all awry, but there was a time when

Corporal Henry Wood was the smartest man in the 117th Foot. We were in India then, in cantonments,* at a place we'll call Bhurtee. Barclay, who died the other day, was sergeant in the same company as myself, and the belle of the regiment—aye, and the finest girl that ever had the breath of life between her lips—was Nancy Devoy, the daughter of the colour-sergeant. There were two men who loved her, and one whom she loved; and you'll smile when you look at this poor thing huddled before the fire, and hear me say that it was for my good looks that she loved me.

'Well, though I had her heart her father was set upon her marrying Barclay. I was a harum-scarum, reckless lad, and he had had an education, and was already marked for the sword-belt. But the girl held true to me, and it seemed that I would have had her, when the Mutiny broke out, and all Hell was loose in the country.

'We were shut up in Bhurtee, the regiment of us, with half a battery of artillery, a company of Sikhs, and a lot of civilians and women-folk. There were ten thousand rebels round us, and they were as keen as a set of terriers round a rat-cage. About the second week of it our water gave out, and it was a question whether we could communicate with General Neill's* column, which was moving up country. It was our only chance, for we could not hope to fight our way out with all the women and children, so I volunteered to go out and warn General Neill of our danger. My offer was accepted, and I talked it over with Sergeant Barclay, who was supposed to know the ground better than any other man, and who drew up a route by which I might get through the rebel lines. At ten o'clock the same night I started off upon my journey. There were a thousand lives to save, but it was of only one that I was thinking when I dropped over the wall that night.

'My way ran down a dried-up watercourse, which we hoped would screen me from the enemy's sentries, but as I crept round the corner of it I walked right into six of them, who were crouching down in the dark waiting for me. In an instant I was stunned with a blow, and bound hand and foot.

But the real blow was to my heart and not to my head, for as I came to and listened to as much as I could understand of their talk, I heard enough to tell me that my comrade, the very man who had arranged the way I was to take, had betrayed me by means of a native servant into the hands of the enemy.

'Well, there's no need for me to dwell on that part of it. You know now what James Barclay was capable of. Bhurtee was relieved by Neill next day, but the rebels took me away with them in their retreat, and it was many a long year before ever I saw a white face again. I was tortured, and tried to get away, and was captured and tortured again. You can see for yourselves the state in which I was left. Some of them that fled into Nepal* took me with them, and then afterwards I was up past Darjeeling*. The hill-folk up there murdered the rebels who had me, and I became their slave for a time until I escaped, but instead of going south I had to go north, until I found myself among the Afghans. There I wandered about for many a year, and at last came back to the Punjab,* where I lived mostly among the natives, and picked up a living by the conjuring tricks that I had learned. What use was it for me, a wretched cripple, to go back to England, or to make myself known to my old comrades? Even my wish for revenge would not make me do that. I had rather that Nancy and my old pals should think of Harry Wood as having died with a straight back, than see him living and crawling with a stick like a chimpanzee. They never doubted that I was dead, and I meant that they never should. I heard that Barclay had married Nancy, and that he was rapidly rising in the regiment, but even that did not make me speak.

'But when one gets old, one has a longing for home. For years I've been dreaming of the bright green fields and the hedges of England. At last I determined to see them before I died. I saved enough to bring me across, and then I came here where the soldiers are, for I know their ways, and how to amuse them, and so earn enough to keep me.'

'Your narrative is most interesting,' said Sherlock Holmes. 'I have already heard of your meeting with Mrs Barclay and your mutual recognition. You then, as I understand, followed her home and saw through the window an altercation between her husband and her, in which she doubtless cast his conduct to you in his teeth. Your own feelings overcame you, and you ranáacross the lawn, and broke in upon them.'

'I did, sir, and at the sight of me he looked as I have never seen a man look before, and over he went with his head on the fender. But he was dead before he fell. I read death on his face as plain as I can read that text over the fire. The bare sight of me was like a bullet through his guilty heart.'

'And then?'

'Then Nancy fainted, and I caught up the key of the door from her hand, intending to unlock it and get help. But as I was doing it it seemed to me better to leave it alone and get away, for the thing might look black against me, and any way my secret would be out if I were taken. In my haste I thrust the key into my pocket, and dropped my stick while I was chasing Teddy, who had run up the curtain. When I got him into his box, from which he had slipped, I was off as fast as I could run.'

'Who's Teddy?' asked Holmes.

The man leaned over and pulled up the front of a kind of hutch in the corner. In an instant out there slipped a beautiful reddish-brown creature, thin, and lithe, with the legs of a stoat, a long thin nose, and a pair of the finest red eyes that ever I saw in an animal's head.

'It's a mongoose!' I cried.

'Well, some call them that, and some call them ichneumon,' said the man. 'Snake-catcher is what I call them, and Teddy is amazing quick on cobras. I have one here without the fangs, and Teddy catches it every night to please the folk in the canteen. Any other point, sir?'

'Well, we may have to apply to you again if Mrs Barclay should prove to be in serious trouble.'

'In that case, of course, I'd come forward.'

'But if not, there is no object in raking up this scandal against a dead man, foully as he has acted. You have, at least, the satisfaction of knowing that for thirty years of his life his conscience bitterly reproached him for his wicked deed. Ah, there goes Major Murphy on the other side of the street. Good-bye, Wood; I want to learn if anything has happened since yesterday.'

We were in time to overtake the Major before he reached the corner.

'Ah, Holmes,' he said, 'I suppose you have heard that all this fuss has come to nothing?'

'What, then?'

'The inquest is just over. The medical evidence showed conclusively that death was due to apoplexy. You see, it was quite a simple case after all.'

'Oh, remarkably superficial,' said Holmes, smiling. 'Come, Watson, I don't think we shall be wanted in Aldershot any more.'

'There's one thing,' said I, as we walked down to the station; 'if the husband's name was James, and the other was Henry, what was this talk about David?'

'That one word, my dear Watson, should have told me the whole story had I been the ideal reasoner which you are so fond of depicting. It was evidently a term of reproach.'

'Of reproach?'

'Yes, David strayed a little now and then, you know, and on one occasion in the same direction as Sergeant James Barclay. You remember the small affair of Uriah and Bathsheba?* My Biblical knowledge is a trifle rusty, I fear, but you will find the story in the first or second of Samuel.'

The Resident Patient

IN glancing over the somewhat incoherent series of memoirs with which I have endeavoured to illustrate a few of the mental peculiarities of my friend, Mr Sherlock Holmes, I have been struck by the difficulty which I have experienced in picking out examples which shall in every way answer my purpose. For in those cases in which Holmes has performed some *tour-de-force** of analytical reasoning, and has demonstrated the value of his peculiar methods of investigation, the facts themselves have often been so slight or so commonplace that I could not feel justified in laying them before the public. On the other hand, it has frequently happened that he has been concerned in some research where the facts have been of the most remarkable and dramatic character, but where the share which he has himself taken in determining their causes has been less pronounced than I, as his biographer, could wish. The small matter which I have chronicled under the heading of 'A Study in Scarlet,' and that other later one connected with the loss of the *Gloria Scott*, may serve as examples of this Scylla and Charybdis* which are for ever threatening his historian. It may be that, in the business of which I am now about to write, the part which my friend played is not sufficiently accentuated; and yet the whole train of circumstances is so remarkable that I cannot bring myself to omit it entirely from this series.

I cannot be sure of the exact date,* for some of my memoranda upon the matter have been mislaid, but it must have been towards the end of the first year during which Holmes and I shared chambers in Baker Street. It was boisterous October weather, and we had both remained indoors all day, I because I feared with my shaken health to face the keen autumn wind, while he was deep in some of those abstruse chemical investigations which absorbed him utterly as long as he was engaged upon them. Towards evening, however, the breaking of a test-tube brought his research to a premature ending, and he sprang up from his chair with an exclamation of impatience and a clouded brow.

'A day's work ruined, Watson,' said he, striding across to the window. 'Ha! the stars are out and the wind has fallen. What do you say to a ramble through London?'

I was weary of our little sitting-room, and gladly acquiesced, muffling myself nose-high* against the keen night air. For three hours we strolled about together, watching the ever-changing kaleidoscope of life as it ebbs and flows through Fleet Street* and the Strand. Holmes had shaken off his temporary ill-humour, and his characteristic talk, with its keen observance of detail and subtle power of inference, held me amused and enthralled. It was ten o'clock before we reached Baker Street again. A brougham was waiting at our door.

'Hum! A doctor's—general practitioner, I perceive,' said Holmes. 'Not been long in practice, or had much to do. Come to consult us, I fancy! Lucky we came back!'

I was sufficiently conversant with Holmes's methods to be able to follow his reasoning, and to see that the nature and state of the various medical instruments in the wicker basket which hung in the lamp-light inside the brougham had given him the data for his swift deduction. The light in our window above showed that this late visit was indeed intended for us. With some curiosity as to what could have sent a brother medico to us at such an hour, I followed Holmes into our sanctum.

A pale, taper-faced man with sandy whiskers rose up from a chair by the fire as we entered. His age may not have been more than three or four and thirty, but his haggard expression and unhealthy hue told of a life which had sapped his strength and robbed him of his youth. His manner was nervous and shy, like that of a sensitive gentleman, and the thin white hand which he laid on the mantelpiece as he rose was that of an artist rather than of a surgeon. His dress was quiet and sombre, a black frock-coat, dark trousers, and a touch of colour about his necktie.

'Good evening, Doctor,' said Holmes, cheerily; 'I am glad to see that you have only been waiting a very few minutes.'

'You spoke to my coachman, then?'

'No, it was the candle on the side-table that told me. Pray resume your seat and let me know how I can serve you.'

'My name is Doctor Percy Trevelyan,' said our visitor, 'and I live at 403, Brook Street.'*

'Are you not the author of a monograph upon obscure nervous lesions?'* I asked.

His pale cheeks flushed with pleasure at hearing that his work was known to me.

'I so seldom hear of the work that I thought it was quite dead,' said he. 'My publishers give me a most discouraging account of its sale. You are yourself, I presume, a medical man?'

'A retired Army surgeon.'

'My own hobby has always been nervous disease. I should wish to make it an absolute specialty, but, of course, a man must take what he can get at first.* This, however, is beside the question, Mr Sherlock Holmes, and I quite appreciate how valuable your time is. The fact is that a very singular train of events has occurred recently at my house in Brook Street, and to-night they came to such a head that I felt it was quite impossible for me to wait another hour before asking for your advice and assistance.'

Sherlock Holmes sat down and lit his pipe. 'You are very welcome to both,' said he. 'Pray let me have a detailed account of what the circumstances are which have disturbed you.'

'One or two of them are so trivial,' said Dr Trevelyan, 'that really I am almost ashamed to mention them. But the matter is so inexplicable, and the recent turn which it has taken is so elaborate, that I shall lay it all before you, and you shall judge what is essential and what is not.

'I am compelled, to begin with, to say something of my own college career. I am a London University man, you know, and I am sure you will not think that I am unduly singing my own praises if I say that my student career was considered by my professors to be a very promising one. After I had graduated I continued to devote myself to research, occupying a minor position in King's College Hospital,* and I was fortunate enough to excite considerable interest by my research into the pathology of catalepsy, and finally to win the Bruce Pinkerton prize and medal by the monograph on nervous lesions to which your friend has just alluded. I should not go too far if I were to say that there was a general impression at that time that a distinguished career lay before me.

'But the one great stumbling-block lay in my want of capital. As you will readily understand, a specialist who aims high is compelled to start in one of a dozen streets in the Cavendish Square quarter,* all of which entail enormous rents and furnishing expenses. Besides this preliminary outlay, he must be prepared to keep himself for some years, and to hire a presentable carriage and horse. To do this was quite beyond my power, and I could only hope that by economy I might in ten years' time save enough to enable me to put up my plate. Suddenly, however, an unexpected incident opened up quite a new prospect to me.

'This was a visit from a gentleman of the name of Blessington, who was a complete stranger to me. He came up into my room one morning, and plunged into business in an instant.

' "You are the same Percy Trevelyan who has had so distinguished a career and won a great prize lately?' said he. I bowed.

' "Answer me frankly," he continued, "for you will find it to your interest to do so. You have all the cleverness which makes a successful man. Have you the tact?"

'I could not help smiling at the abruptness of the question.

' "I trust that I have my share," I said.

' "Any bad habits? Not drawn towards drink, eh?"

' "Really, sir!" I cried.

' "Quite right! That's all right! But I was bound to ask. With all these qualities why are you not in practice?"

'I shrugged my shoulders.

' "Come, come!" said he, in his bustling way. "It's the old story. More in your brains than in your pocket, eh? What would you say if I were to start you in Brook Street?"

'I stared at him in astonishment.

' "Oh, it's for my sake, not for yours," he cried. "I'll be perfectly frank with you, and if it suits you it will suit me very well. I have a few thousands to invest, d'ye see, and I think I'll sink them in you."

' "But why?" I gasped.

' "Well, it's just like any other speculation, and safer than most."

' "What am I to do, then?"

' "I'll tell you. I'll take the house, furnish it, pay the maids, and run the whole place. All you have to do is just to wear out your chair in the consulting-room. I'll let you have pocket-money and everything. Then you hand over to me three-quarters of what you earn and you keep the other quarter for yourself."

'This was the strange proposal, Mr Holmes, with which the man Blessington approached me. I won't weary you with the account of how we bargained and negotiated. It ended in my moving into the house next Lady Day* and starting in practice on very much the same conditions as he had suggested. He came himself to live with me in the character of a resident patient.* His heart was weak, it appears, and he needed constant medical supervision. He turned the two best rooms on the first floor into a sitting-room and bed-room for himself. He was a man of singular habits, shunning company and very seldom going out. His life was irregular, but in one respect he was regularity itself. Every evening at the same hour he walked into the consulting-room, examined the books, put down five and threepence for every guinea* that I had earned, and carried the rest off to the strong box in his own room.

'I may say with confidence that he never had occasion to regret his speculation. From the first it was a success. A few good cases and the reputation which I had won in the hospital brought me rapidly to the front, and during the last few years I have made him a rich man.

'So much, Mr Holmes, for my past history and for my relations with Mr Blessington. It only remains for me now to tell you what has occurred to bring me here to-night.

'Some weeks ago Mr Blessington came down to me in, as it seemed to me, a state of considerable agitation. He spoke of some burglary which, he said, had been committed in the West-end,* and he appeared, I remember, to be quite unnecessarily excited about it, declaring that a day should not pass before we should add stronger bolts to our windows and doors. For a week he continued to be in quite a peculiar state of restlessness, peering continually out of the windows, and ceasing to take the short walk which had usually been the prelude to his dinner. From his manner it struck me that he was in mortal dread of something or somebody, but when I questioned him upon the point he became so offensive that I was compelled to drop the subject. Gradually as time passed his fears appeared to die away, and he had renewed his former habits, when a fresh event reduced him to the pitiable state of prostration in which he now lies.

'What happened was this. Two days ago I received the letter which I now read to you. Neither address nor date is attached to it.

' "A Russian nobleman who is now resident in England," it runs, "would be glad to avail himself of the professional assistance of Dr Percy Trevelyan. He has been for some years a victim to cataleptic attacks, on which, as is well known, Dr Trevelyan is an authority. He proposes to call at about a quarter-past six tomorrow evening, if Dr Trevelyan will make it convenient to be at home."

'This letter interested me deeply, because the chief difficulty in the study of catalepsy is the rareness of the disease. You may believe, then, that I was in my consulting-room when, at the appointed hour, the page showed in the patient.

'He was an elderly man, thin, demure, and common-place—by no means the conception one forms of a Russian nobleman. I was much more struck by the appearance of his companion. This was a tall young man, surprisingly hand-some, with a dark, fierce face, and the limbs and chest of a Hercules.* He had his hand under the other's arm as they entered, and helped him to a chair with a tenderness which one would hardly have expected from his appearance.

' "You will excuse my coming in, Doctor," said he to me, speaking English with a slight lisp. 'This is my father, and his health is a matter of the most overwhelming importance to me.'

'I was touched by this filial anxiety. "You would, perhaps, care to remain during the consultation," said I.

' "Not for the world," he cried, with a gesture of horror. "It is more painful to me than I can express. If I were to see my father in one of those dreadful seizures, I am convinced that I should never survive it. My own nervous system is an exceptionally sensitive one. With your permission I will remain in the waiting-room while you go into my father's case.'

'To this, of course, I assented, and the young man withdrew. The patient and I then plunged into a discussion of his case, of which I took exhaustive notes. He was not remarkable for intelligence, and his answers were frequently obscure, which I attributed to his limited acquaintance with our language. Suddenly, however, as I sat writing he ceased to give any answer at all to my inquiries, and on my turning towards him I was shocked to see that he was sitting bolt upright in his chair, staring at me with a perfectly blank and rigid face. He was again in the grip of his mysterious malady.

'My first feeling, as I have just said, was one of pity and horror. My second, I fear, was rather one of professional satisfaction. I made notes of my patient's pulse and tempera-ture, tested the rigidity of his muscles, and examined his reflexes. There was nothing markedly abnormal in any of these conditions, which harmonized with my former experi-ences. I had obtained good results in such cases by the inhalation of nitrite of amyl,* and the present seemed an admirable opportunity of testing its virtues. The bottle was downstairs in my laboratory, so, leaving my patient seated in his chair, I ran down to get it. There was some little delay in finding it—five minutes, let us say—and then I returned. Imagine my amazement to find the room empty and the patient gone!

'Of course, my first act was to run into the waiting-room. The son had gone also. The hall door had been closed, but not shut. My page who admits patients is a new boy, and by no means quick. He waits downstairs, and runs up to show patients out when I ring the consulting-room bell. He had heard nothing, and the affair remained a complete mystery. Mr Blessington came in from his walk shortly afterwards, but I did not say anything to him upon the subject, for, to tell the truth, I have got in the way of late of holding as little communication with him as possible.

'Well, I never thought that I should see anything more of the Russian and his son, so you can imagine my amazement when at the very same hour this evening they both came marching into my consulting-room, just as they had done before.

' "I feel that I owe you a great many apologies for my abrupt departure yesterday, Doctor," said my patient.

' "I confess that I was very much surprised at it," said I.

' "Well, the fact is," he remarked, "that when I recover from these attacks my mind is always very clouded as to all that has gone before. I woke up in a strange room, as it seemed to me, and made my way out into the street in a sort of dazed way when you were absent.

' "And I," said the son, "seeing my father pass the door of the waiting-room, naturally thought that the consultation had come to an end. It was not until we had reached home that I began to realize the true state of affairs."

' "Well," said I, laughing, "there is no harm done, except that you puzzled me terribly; so if you, sir, would kindly step into the waiting-room, I shall be happy to continue our consultation, which was brought to so abrupt an ending."

'For half an hour or so I discussed the old gentleman's symptoms with him, and then, having prescribed for him, I saw him go off on the arm of his son.

'I have told you that Mr Blessington generally chose this hour of the day for his exercise. He came in shortly afterwards and passed upstairs. An instant later I heard him running down, and he burst into my consulting-room like a man who is mad with panic.

' "Who has been in my room?" he cried.

' "No one," said I.

' "It's a lie!" he yelled. "Come up and look."

'I passed over the grossness of his language, as he seemed half out of his mind with fear. When I went upstairs with him he pointed to several footprints upon the light carpet.

' "D'you mean to say those are mine?" he cried.

'They were certainly very much larger than any which he could have made, and were evidently quite fresh. It rained hard this afternoon, as you know, and my patients were the only people who called. It must have been the case, then, that the man in the waiting-room had for some unknown reason, while I was busy with the other, ascended to the room of my resident patient. Nothing had been touched or taken, but there were the footprints to prove that the intrusion was an undoubted fact.

'Mr Blessington seemed more excited over the matter than I should have thought possible, though, of course, it was enough to disturb anybody's peace of mind. He actually sat crying in an arm-chair, and I could hardly get him to speak coherently. It was his suggestion that I should come round to you, and of course I at once saw the propriety of it, for certainly the incident is a very singular one, though he appears to completely overrate its importance. If you would only come back with me in my brougham,* you would at least be able to soothe him, though I can hardly hope that you will be able to explain this remarkable occurrence.'

Sherlock Holmes had listened to this long narrative with an intentness which showed me that his interest was keenly aroused. His face was as impassive as ever, but his lids had drooped more heavily over his eyes, and his smoke had curled up more thickly from his pipe to emphasize each curious episode in the doctor's tale. As our visitor concluded Holmes sprang up without a word, handed me my hat, picked up his own from the table, and followed Dr Trevelyan to the door. Within a quarter of an hour we had been dropped at the door of the physician's residence in Brook Street, one of those sombre, flat-faced houses which one associates with a West-end practice. A small page admitted us, and we began at once to ascend the broad, well-carpeted stair.

But a singular interruption brought us to a standstill. The light at the top was suddenly whisked out, and from the darkness came a reedy, quavering voice.

'I have a pistol,' it cried; 'I give you my word that I'll fire if you come any nearer.'

'This really grows outrageous, Mr Blessington,' cried Dr Trevelyan.

'Oh, then it is you, Doctor?' said the voice, with a great heave of relief. 'But those other gentlemen, are they what they pretend to be?'

We were conscious of a long scrutiny out of the darkness.

'Yes, yes, it's all right,' said the voice at last. 'You can come up, and I am sorry if my precautions have annoyed you.'

He re-lit the stair gas as he spoke, and we saw before us a singular-looking man, whose appearance, as well as his voice, testified to his jangled nerves. He was very fat, but had apparently at some time been much fatter, so that the skin hung about his face in loose pouches, like the cheeks of a blood-hound. He was of a sickly colour, and his thin sandy hair seemed to bristle up with the intensity of his emotion. In his hand he held a pistol, but he thrust it into his pocket as we advanced.

'Good evening, Mr Holmes,' said he; 'I am sure I am very much obliged to you for coming round. No one ever needed your advice more than I do. I suppose that Dr Trevelyan has told you of this most unwarrantable intrusion into my rooms.'

'Quite so,' said Holmes. 'Who are these two men, Mr Blessington, and why do they wish to molest you?'

'Well, well,' said the resident patient, in a nervous fashion, 'of course, it is hard to say that. You can hardly expect me to answer that, Mr Holmes.'

'Do you mean that you don't know?'

'Come in here, if you please. Just have the kindness to step in here.'

He led the way into his bedroom, which was large and comfortably furnished.

'You see that,' said he, pointing to a big black box at the end of his bed. 'I have never been a very rich man, Mr Holmes—never made but one investment in my life, as Dr Trevelyan would tell you. But I don't believe in bankers. I would never trust a banker, Mr Holmes. Between ourselves, what little I have is in that box, so you can understand what it means to me when unknown people force themselves into my rooms.'

Holmes looked at Blessington in his questioning way, and shook his head.

'I cannot possibly advise you if you try to deceive me,' said he.

'But I have told you everything.'

Holmes turned on his heel with a gesture of disgust. 'Good-night, Dr Trevelyan,' said he.

'And no advice for me?' cried Blessington, in a breaking voice.

'My advice to you, sir, is to speak the truth.'

A minute later we were in the street and walking for home. We had crossed Oxford Street,* and were half-way down Harley Street* before I could get a word from my companion.

'Sorry to bring you out on such a fool's errand, Watson,' he said at last. 'It is an interesting case, too, at the bottom of it.'

'I can make little of it,' I confessed.

'Well, it is quite evident that there are two men—more, perhaps, but at least two—who are determined for some reason to get at this fellow Blessington. I have no doubt in my mind that both on the first and on the second occasion that young man penetrated to Blessington's room, while his confederate, by an ingenious device, kept the doctor from interfering.'

'And the catalepsy!'

'A fraudulent imitation, Watson, though I should hardly dare to hint as much to our specialist. It is a very easy complaint to imitate. I have done it myself.'

'And then?'

'By the purest chance Blessington was out on each occasion. Their reason for choosing so unusual an hour for a consultation was obviously to insure that there should be no other patient in the waiting-room. It just happened, however, that this hour coincided with Blessington's constitutional, which seems to show that they were not very well acquainted with his daily routine. Of course, if they had been merely after plunder they would at least have made some attempt to search for it. Besides, I can read in a man's eye when it is his own skin that he is frightened for. It is inconceivable that this fellow could have made two such vindictive enemies as these appear to be without knowing of it. I hold it, therefore, to be certain that he does know who these men are, and that for reasons of his own he suppresses it. It is just possible that to-morrow may find him in a more communicative mood.'

'Is there not one alternative,' I suggested, 'grotesquely improbable, no doubt, but still just conceivable? Might the whole story of the cataleptic Russian and his son be a concoction of Dr Trevelyan's, who has, for his own purposes, been in Blessington's rooms?'

I saw in the gaslight that Holmes wore an amused smile at this brilliant departure of mine.

'My dear fellow,' said he, 'it was one of the first solutions which occurred to me, but I was soon able to corroborate the doctor's tale. This young man has left prints upon the stair carpet which made it quite superfluous for me to ask to see those which he had made in the room. When I tell you that his shoes were square-toed, instead of being pointed like Blessington's, and were quite an inch and a third longer than the doctor's, you will acknowledge that there can be no doubt as to his individuality. But we may sleep on it now, for I shall be surprised if we do not hear something further from Brook Street in the morning.'

Sherlock Holmes's prophecy was soon fulfilled, and in a dramatic fashion. At half-past seven next morning, in the first dim glimmer of daylight, I found him standing by my bedside in his dressing-gown.

'There's a brougham waiting for us, Watson,' said he.

'What's the matter, then?'

'The Brook Street business.'

'Any fresh news?'

'Tragic but ambiguous,' said he, pulling up the blind. 'Look at this—a sheet from a notebook with "For God's sake, come at once—P. T." scrawled upon it in pencil. Our friend the doctor was hard put to it when he wrote this. Come along, my dear fellow, for it's an urgent call.'

In a quarter of an hour or so we were back at the physician's house. He came running out to meet us with a face of horror.

'Oh, such a business!' he cried, with his hands to his temples.

'What, then?'

'Blessington has committed suicide!'

Holmes whistled.

'Yes, he hanged himself during the night.'

We had entered, and the doctor had preceded us into what was evidently his waiting-room.

'I really hardly know what I am doing,' he cried. 'The police are already upstairs. It has shaken me most dreadfully.'

'When did you find it out?'

'He has a cup of tea taken in to him early every morning. When the maid entered about seven, there the unfortunate fellow was hanging in the middle of the room. He had tied his cord to the hook on which the heavy lamp used to hang, and he had jumped off from the top of the very box that he showed us yesterday.'

Holmes stood for a moment in deep thought.

'With your permission,' said he at last, 'I should like to go upstairs and look into the matter.' We both ascended, followed by the doctor.

It was a dreadful sight which met us as we entered the bedroom door. I have spoken of the impression of flabbiness which this man Blessington conveyed. As he dangled from the hook it was exaggerated and intensified until he was scarce human in his appearance. The neck was drawn out like a plucked chicken's, making the rest of him seem the more obese and unnatural by the contrast. He was clad only in his long night-dress, and his swollen ankles and ungainly feet protruded starkly from beneath it. Beside him stood a smart-looking police inspector, who was taking notes in a pocket-book.

'Ah, Mr Holmes,' said he, heartily, as my friend entered. 'I am delighted to see you.'

'Good morning, Lanner,' answered Holmes. 'You won't think me an intruder, I am sure. Have you heard of the events which led up to this affair?'

'Yes, I heard something of them.'

'Have you formed any opinion?'

'As far as I can see, the man has been driven out of his senses by fright. The bed has been well slept in, you see. There's his impression deep enough. It's about five in the morning, you know, that suicides are most common. That would be about his time for hanging himself. It seems to have been a very deliberate affair.'

'I should say that he has been dead about three hours, judging by the rigidity of the muscles,' said I.

'Noticed anything peculiar about the room?' asked Holmes.

'Found a screwdriver and some screws on the wash-hand stand. Seems to have smoked heavily during the night, too. Here are four cigar ends that I picked out of the fireplace.'

'Hum!' said Holmes. 'Have you got his cigar-holder?'

'No, I have seen none.'

'His cigar-case, then?'

'Yes, it was in his coat pocket.'

Holmes opened it and smelled the single cigar which it contained.

'Oh, this is a Havana, and these others are cigars of the peculiar sort which are imported by the Dutch from their East Indian colonies. They are usually wrapped in straw, you know, and are thinner for their length than any other brand.' He picked up the four ends and examined them with his pocket lens.

'Two of these have been smoked from a holder and two without,' said he. 'Two have been cut by a not very sharp knife, and two have had the ends bitten off by a set of excellent teeth. This is no suicide, Mr Lanner. It is a very deeply-planned and cold-blooded murder.'

'Impossible!' cried the inspector.

'And why?'

'Why should anyone murder a man in so clumsy a fashion as by hanging him?'

'That is what we have to find out.'

'How could they get in?'

'Through the front door.'

'It was barred in the morning.'

'Then it was barred after them.'

'How do you know?'

'I saw their traces. Excuse me a moment, and I may be able to give you some further information about it.'

He went over to the door, and turning the lock he examined it in his methodical fashion. Then he took out the key, which was on the inside, and inspected that also. The bed, the carpet, the chairs, the mantelpiece, the dead body, and the rope were each in turn examined, until at last he professed himself satisfied, and with my aid and that of the inspector cut down the wretched object, and laid it reverently under a sheet.

'How about this rope?' he asked.

'It is cut off this,' said Dr Trevelyan, drawing a large coil from under the bed. 'He was morbidly nervous of fire, and always kept this beside him, so that he might escape by the window in case the stairs were burning.'

'That must have saved them trouble,' said Holmes, thoughtfully. 'Yes, the actual facts are very plain, and I shall be surprised if by the afternoon I cannot give you the reasons for them as well. I will take this photograph of Blessington which I see upon the mantelpiece, as it may help me in my inquiries.'

'But you have told us nothing,' cried the doctor.

'Oh, there can be no doubt as to the sequence of events,' said Holmes. 'There were three of them in it: the young man, the old man, and a third to whose identity I have no clue. The first two, I need hardly remark, are the same who masqueraded as the Russian Count and his son, so we can give a very full description of them. They were admitted by a confederate inside the house. If I might offer you a word of advice, Inspector, it would be to arrest the page, who, as I understand, has only recently come into your service, Doctor.'

'The young imp cannot be found,' said Dr Trevelyan; 'the maid and the cook have just been searching for him.'

Holmes shrugged his shoulders.

'He has played a not unimportant part in this drama,' said he. 'The three men having ascended the stair, which they did on tiptoe, the elder man first, the younger man second, and the unknown man in the rear—'

'My dear Holmes!' I ejaculated.

'Oh, there could be no question as to the superimposing of the footmarks. I had the advantage of learning which was which last night. They ascended then to Mr Blessington's room, the door of which they found to be locked. With the help of a wire, however, they forced round the key. Even without a lens, you will perceive by the scratches on this ward where the pressure was applied.

'On entering the room, their first proceeding must have been to gag Mr Blessington. He may have been asleep, or he may have been so paralysed with terror as to have been unable to cry out. These walls are thick, and it is conceivable that his shriek, if he had time to utter one, was unheard.

'Having secured him, it is evident to me that a consultation of some sort was held. Probably it was something in the nature of a judicial proceeding. It must have lasted for some time, for it was then that these cigars were smoked. The older man sat in that wicker chair: it was he who used the cigar-holder. The younger man sat over yonder: he knocked his ash off against the chest of drawers. The third fellow paced up and down. Blessington, I think, sat upright in the bed, but of that I cannot be absolutely certain.

'Well, it ended by their taking Blessington and hanging him. The matter was so prearranged that it is my belief that they brought with them some sort of block or pulley which might serve as a gallows. That screwdriver and those screws were, as I conceive, for fixing it up. Seeing the hook, however, they naturally saved themselves the trouble. Having finished their work they made off, and the door was barred behind them by their confederate.'

We had all listened with the deepest interest to this sketch of the night's doings, which Holmes had deduced from signs so subtle and minute, that even when he had pointed them out to us, we could scarcely follow him in his reasonings. The inspector hurried away on the instant to make inquiries about the page, while Holmes and I returned to Baker Street for breakfast.

'I'll be back by three,' said he when we had finished our meal. 'Both the inspector and the doctor will meet me here at that hour, and I hope by that time to have cleared up any little obscurity which the case may still present.'

Our visitors arrived at the appointed time, but it was a quarter to four before my friend put in an appearance. From his expression as he entered, however, I could see that all had gone well with him.

'Any news, Inspector?'

'We have got the boy, sir.'

'Excellent, and I have got the men.'

'You have got them!' we cried all three.

'Well, at least I have got their identity. This so-called Blessington is, as I expected, well known at headquarters, and so are his assailants. Their names are Biddle, Hayward, and Moffat.'

'The Worthingdon bank gang,' cried the inspector.

'Precisely,' said Holmes.

'Then Blessington must have been Sutton?'

'Exactly,' said Holmes.

'Why, that makes it as clear as crystal,' said the inspector. But Trevelyan and I looked at each other in bewilderment.

'You must surely remember the great Worthingdon bank business,' said Holmes; 'five men were in it, these four and a fifth called Cartwright. Tobin, the caretaker, was murdered, and the thieves got away with seven thousand pounds. This was in 1875. They were all five arrested, but the evidence against them was by no means conclusive. This Blessington, or Sutton, who was the worst of the gang, turned informer. On his evidence, Cartwright was hanged and the other three got fifteen years apiece. When they got out the other day, which was some years before their full term,* they set themselves, as you perceive, to hunt down the traitor and to avenge the death of their comrade upon him. Twice they tried to get at him and failed; a third time, you see, it came off. Is there anything further which I can explain, Dr Trevelyan?'

'I think you have made it all remarkably clear,' said the doctor. 'No doubt the day on which he was so perturbed was the day when he had read of their release in the newspapers.'

'Quite so. His talk about a burglary was the merest blind.'

'But why could he not tell you this?'

'Well, my dear sir, knowing the vindictive character of his old associates, he was trying to hide his own identity from everybody as long as he could. His secret was a shameful one, and he could not bring himself to divulge it. However, wretch as he was, he was still living under the shield of British law, and I have no doubt, Inspector, that you will see that, though that shield may fail to guard, the sword of justice is still there to avenge.'

Such were the singular circumstances in connection with the resident patient and the Brook Street doctor. From that night nothing has been seen of the three murderers by the police, and it is surmised at Scotland Yard that they were among the passengers of the ill-fated steamer *Norah Creina*,* which was lost some years ago with all hands upon the Portuguese coast, some leagues to the north of Oporto. The proceedings against the page broke down for want of evidence, and the 'Brook Street Mystery,' as it was called, has never, until now, been fully dealt with in any public print.

The Greek Interpreter

DURING my long and intimate acquaintance with Mr Sherlock Holmes I had never heard him refer to his relations, and hardly ever to his own early life. This reticence upon his part had increased the somewhat inhuman effect which he produced upon me, until sometimes I found myself regarding him as an isolated phenomenon, a brain without a heart, as deficient in human sympathy as he was pre-eminent in intelligence. His aversion to women, and his disinclination to form new friendships, were both typical of his unemotional character, but not more so than his complete suppression of every reference to his own people. I had come to believe that he was an orphan with no relatives living, but one day, to my very great surprise, he began to talk to me about his brother.

It was after tea on a summer evening, and the conversation, which had roamed in a desultory, spasmodic fashion from golf clubs to the causes of the change in the obliquity of the ecliptic,* came round at last to the question of atavism* and hereditary aptitudes. The point under discussion was how far any singular gift in an individual was due to his ancestry, and how far to his own early training.

'In your own case,' said I, 'from all that you have told me it seems obvious that your faculty of observation and your peculiar facility for deduction are due to your own systematic training.'

'To some extent,' he answered thoughtfully. 'My ancestors were country squires, who appear to have led much the same life as is natural to their class. But, none the less, my turn that way is in my veins, and may have come with my grandmother, who was the sister of Vernet,* the French artist. Art in the blood is liable to take the strangest forms.'

'But how do you know that it is hereditary?'

'Because my brother Mycroft possesses it in a larger degree than I do.'

This was news to me, indeed. If there were another man with such singular powers in England, how was it that neither police nor public had heard of him? I put the question, with a hint that it was my companion's modesty which made him acknowledge his brother as his superior. Holmes laughed at my suggestion.

'My dear Watson,' said he, 'I cannot agree with those who rank modesty among the virtues. To the logician all things should be seen exactly as they are, and to under-estimate oneself is as much a departure from truth as to exaggerate one's own powers. When I say, therefore, that Mycroft has better powers of observation than I, you may take it that I am speaking the exact and literal truth.'

'Is he your junior?'

'Seven years my senior.'

'How comes it that he is unknown?'

'Oh, he is very well known in his own circle.'

'Where, then?'

'Well, in the Diogenes Club,* for example.'

I had never heard of the institution, and my face must have proclaimed as much, for Sherlock Holmes pulled out his watch.

'The Diogenes Club is the queerest club in London, and Mycroft, one of the queerest men. He's always there from a quarter to five till twenty to eight. It's six now, so if you care for a stroll this beautiful evening I shall be very happy to introduce you to two curiosities.'

Five minutes later we were in the street, walking towards Regent Circus.*

'You wonder,' said my companion, 'why it is that Mycroft does not use his powers for detective work. He is incapable of it.'

'But I thought you said—!'

'I said that he was my superior in observation and deduction. If the art of the detective began and ended in reasoning from an arm-chair, my brother would be the greatest criminal agent that ever lived. But he has no ambition and no energy. He will not even go out of his way to verify his own solutions, and would rather be considered wrong than take the trouble to prove himself right. Again and again I have taken a problem to him, and have received an explanation which has afterwards proved to be the correct one. And yet he was absolutely incapable of working out the practical points which must be gone into before a case could be laid before a judge or jury.'

'It is not his profession, then?'

'By no means. What is to me a means of livelihood is to him the merest hobby of a dilettante. He has an extraordinary faculty for figures, and audits the books in some of the Government departments. Mycroft lodges in Pall Mall,* and he walks round the corner into Whitehall every morning and back every evening. From year's end to year's end he takes no other exercise, and is seen nowhere else, except only in the Diogenes Club, which is just opposite his rooms.'

'I cannot recall the name.'

'Very likely not. There are many men in London, you know, who, some from shyness, some from misanthropy, have no wish for the company of their fellows. Yet they are not averse to comfortable chairs and the latest periodicals. It is for the convenience of these that the Diogenes Club was started, and it now contains the most unsociable and un-clubbable men in town. No member is permitted to take the least notice of any other one. Save in the Strangers' Room, no talking is, under any circumstances, permitted, and three offences, if brought to the notice of the committee, render the talker liable to expulsion. My brother was one of the founders, and I have myself found it a very soothing atmosphere.'

We had reached Pall Mall as we talked, and were walking down it from the St James's end. Sherlock Holmes stopped at a door some little distance from the Carlton,* and, cautioning me not to speak, he led the way into the hall. Through the glass panelling I caught a glimpse of a large and luxurious room in which a considerable number of men were sitting about and reading papers, each in his own little nook. Holmes showed me into a small chamber which looked out on to Pall Mall, and then, leaving me for a minute, he came back with a companion who I knew could only be his brother.

Mycroft Holmes was a much larger and stouter man than Sherlock. His body was absolutely corpulent, but his face, though massive, had preserved something of the sharpness of expression which was so remarkable in that of his brother. His eyes, which were of a peculiarly light watery grey, seemed to always retain that far-away, introspective look which I had only observed in Sherlock's when he was exerting his full powers.

'I am glad to meet you, sir,' said he, putting out a broad, fat hand, like the flipper of a seal. 'I hear of Sherlock everywhere since you became his chronicler. By the way, Sherlock, I expected to see you round last week to consult me over that Manor House case. I thought you might be a little out of your depth.'

'No, I solved it,' said my friend, smiling.

'It was Adams, of course?'

'Yes, it was Adams.'

'I was sure of it from the first.' The two sat down together in the bow-window of the club. 'To anyone who wishes to study mankind this is the spot,' said Mycroft. 'Look at the magnificent types! Look at these two men who are coming towards us, for example.'

'The billiard-marker and the other?'

'Precisely. What do you make of the other?'

The two men had stopped opposite the window. Some chalk marks over the waistcoat pocket were the only signs of billiards which I could see in one of them. The other was a very small, dark fellow, with his hat pushed back and several packages under his arm.

'An old soldier, I perceive,*' said Sherlock.

'And very recently discharged,' remarked the brother.

'Served in India, I see.'

'And a non-commissioned officer.'

'Royal Artillery, I fancy,' said Sherlock.

'And a widower.'

'But with a child.'

'Children, my dear boy, children.'

'Come,' said I, laughing, 'this is a little too much.'

'Surely,' answered Holmes, 'it is not hard to say that a man with that bearing, expression of authority, and sun-baked skin is a soldier, is more than a private, and is not long from India.'

'That he has not left the service long is shown by his still wearing his "ammunition boots", as they are called,' observed Mycroft.

'He has not the cavalry stride, yet he wore his hat on one side, as is shown by the lighter skin on that side of his brow. His weight is against his being a sapper.* He is in the artillery.'

'Then, of course, his complete mourning shows that he has lost someone very dear. The fact that he is doing his own shopping looks as though it were his wife. He has been buying things for children, you perceive. There is a rattle, which shows that one of them is very young. The wife probably died in child-bed. The fact that he has a picture-book under his arm shows that there is another child to be thought of.'

I began to understand what my friend meant when he said that his brother possessed even keener faculties than he did himself. He glanced across at me, and smiled. Mycroft took snuff from a tortoiseshell box and brushed away the wandering grains from his coat with a large, red silk handkerchief.

'By the way, Sherlock,' said he, 'I have had something quite after your own heart—a most singular problem—submitted to my judgment. I really had not the energy to follow it up, save in a very incomplete fashion, but it gave me a basis for some very pleasing speculations. If you would care to hear the facts—'

'My dear Mycroft, I should be delighted.'

The brother scribbled a note upon a leaf of his pocket-book, and, ringing the bell, he handed it to the waiter.

'I have asked Mr Melas* to step across,' said he. 'He lodges on the floor above me, and I have some slight acquaintance with him, which led him to come to me in his perplexity. Mr Melas is a Greek by extraction, as I understand, and he is a remarkable linguist. He earns his living partly as interpreter in the law courts, and partly by acting as guide to any wealthy Orientals who may visit the Northumberland Avenue* hotels. I think I will leave him to tell his own very remarkable experience in his own fashion.'

A few minutes later we were joined by a short, stout man, whose olive face and coal-black hair proclaimed his southern origin, though his speech was that of an educated Englishman. He shook hands eagerly with Sherlock Holmes, and his dark eyes sparkled with pleasure when he understood that the specialist was anxious to hear his story.

'I do not believe that the police credit me—on my word I do not,' said he, in a wailing voice. 'Just because they have never heard of it before, they think that such a thing cannot be. But I know that I shall never be easy in my mind until I know what has become of my poor man with the sticking-plaster upon his face.'

'I am all attention,' said Sherlock Holmes.

'This is Wednesday evening,' said Mr Melas; 'well, then, it was on Monday night—only two days ago, you understand—that all this happened. I am an interpreter, as, perhaps, my neighbour there has told you. I interpret all languages—or nearly all—but as I am a Greek by birth, and with a Grecian name, it is with that particular tongue that I am principally associated. For many years I have been the chief Greek interpreter in London, and my name is very well known in the hotels.

'It happens, not infrequently, that I am sent for at strange hours, by foreigners who get into difficulties, or by travellers who arrive late and wish my services. I was not surprised, therefore, on Monday night when a Mr Latimer, a very fashionably-dressed young man, came up to my rooms and asked me to accompany him in a cab, which was waiting at the door. A Greek friend had come to see him upon business, he said, and, as he could speak nothing but his own tongue, the services of an interpreter were indispensable. He gave me to understand that his house was some little distance off, in Kensington, and he seemed to be in a great hurry, bustling me rapidly into the cab when we had descended into the street.

'I say into the cab, but I soon became doubtful as to whether it was not a carriage in which I had found myself. It was certainly more roomy than the ordinary four-wheeled disgrace to London, and the fittings, though frayed, were of rich quality. Mr Latimer seated himself opposite to me, and we started off through Charing Cross and up the Shaftesbury Avenue. We had come out upon Oxford Street, and I had ventured some remark as to this being a roundabout way to Kensington, when my words were arrested by the extraordinary conduct of my companion.

'He began by drawing a most formidable-looking bludgeon loaded with lead from his pocket, and switched it backwards and forwards several times, as if to test its weight and strength. Then he placed it, without a word, upon the seat beside him. Having done this, he drew up the windows on each side, and I found to my astonishment that they were covered with paper so as to prevent my seeing through them.

' "I am sorry to cut off your view, Mr Melas," said he. "The fact is that I have no intention that you should see what the place is to which we are driving. It might possibly be inconvenient to me if you could find your way there again."

'As you can imagine, I was utterly taken aback by such an address. My companion was a powerful, broad-shouldered young fellow, and, apart from the weapon, I should not have had the slightest chance in a struggle with him.

' "This is very extraordinary conduct, Mr Latimer," I stammered. "You must be aware that what you are doing is quite illegal."

' "It is somewhat of a liberty, no doubt," said he, "but we'll make it up to you. But I must warn you, however, Mr Melas, that if at any time to-night you attempt to raise an alarm or do anything which is against my interests, you will find it a very serious thing. I beg you to remember that no one knows where you are, and that whether you are in this carriage or in my house, you are equally in my power."

'His words were quiet, but he had a rasping way of saying them which was very menacing. I sat in silence, wondering what on earth could be his reason for kidnapping me in this extraordinary fashion. Whatever it might be, it was perfectly clear that there was no possible use in my resisting, and that I could only wait to see what might befall.

'For nearly two hours we drove without my having the least clue as to where we were going. Sometimes the rattle of the stones told of a paved causeway, and at others our smooth, silent course suggested asphalt,* but save this varia- tion* in sound there was nothing at all which could in the remotest way help me to form a guess as to where we were. The paper over each window was impenetrable to light, and a blue curtain was drawn across the glass-work in front. It was a quarter past seven when we left Pall Mall, and my watch showed me that it was ten minutes to nine when we at last came to a standstill. My companion let down the window and I caught a glimpse of a low, arched doorway with a lamp burning above it. As I was hurried from the carriage it swung open, and I found myself inside the house, with a vague impression of a lawn and trees on each side of me as I entered. Whether these were private grounds, however, or *bona-fide* country was more than I could possibly venture to say.

'There was a coloured gas-lamp inside, which was turned so low that I could see little save that the hall was of some size and hung with pictures. In the dim light I could make out that the person who had opened the door was a small, mean-looking, middle-aged man with rounded shoulders. As he turned towards us the glint of the light showed me that he was wearing glasses.

' "Is this Mr Melas, Harold?" said he.

' "Yes."

' "Well done! Well done! No ill-will, Mr Melas, I hope, but we could not get on without you. If you deal fair with us you'll not regret it; but if you try any tricks, God help you!"

'He spoke in a jerky, nervous fashion, and with some giggling laughs in between, but somehow he impressed me with fear more than the other.

' "What do you want with me?" I asked.

' "Only to ask a few questions of a Greek gentleman who is visiting us, and to let us have the answers. But say no more than you are told to say, or"—here came the nervous giggle again—"you had better never have been born."

'As he spoke he opened a door and showed the way into a room which appeared to be very richly furnished—but again the only light was afforded by a single lamp half turned down. The chamber was certainly large, and the way in which my feet sank into the carpet as I stepped across it told me of its richness. I caught glimpses of velvet chairs, a high, white marble mantelpiece, and what seemed to be a suit of Japanese armour at one side of it. There was a chair just under the lamp, and the elderly man motioned that I should sit in it. The younger had left us, but he suddenly returned through another door, leading with him a gentleman clad in some sort of loose dressing-gown, who moved slowly towards us. As he came into the circle of dim light which enabled me to see him more clearly, I was thrilled with horror at his appearance. He was deadly pale and terribly emaciated, with the protruding, brilliant eyes of a man whose spirit is greater than his strength. But what shocked me more than any signs of physical weakness was that his face was grotesquely criss-crossed with sticking-plaster, and that one large pad of it was fastened over his mouth.

' "Have you the slate, Harold?" cried the older man, as this strange being fell rather than sat down into a chair. "Are his hands loose? Now then, give him the pencil. You are to ask the questions, Mr Melas, and he will write the answers. Ask him first of all whether he is prepared to sign the papers."

'The man's eyes flashed fire.

' "Never," he wrote in Greek upon the slate.

' "On no conditions?" I asked at the bidding of our tyrant.

' "Only if I see her married in my presence by a Greek priest whom I know."

'The man giggled in his venomous way.

' "You know what awaits you, then?"

' "I care nothing for myself."

'These are samples of the questions and answers which made up our strange, half-spoken, half-written conversation. Again and again I had to ask him whether he would give in and sign the document. Again and again I had the same indignant reply. But soon a happy thought came to me. I took to adding on little sentences of my own to each question—innocent ones at first, to test whether either of our companions knew anything of the matter, and then, as I found that they showed no sign, I played a more dangerous game. Our conversation ran something like this:

' "You can do no good by this obstinacy. *Who are you?*"

' "I care not. *I am a stranger in London.*"

' "Your fate will be on your own head. *How long have you been here?*"

' "Let it be so. *Three weeks.*"

' "The property can never be yours. *What ails you?*"

' "It shall not go to villains. *They are starving me.*"

' "You shall go free if you sign. *What house is this?*"

' "I will never sign. *I do not know.*"

' "You are not doing her any service. *What is your name?*"

' "Let me hear her say so. *Kratides.*"

' "You shall see her if you sign. *Where are you from?*"

' "Then I shall never see her. *Athens.*"

'Another five minutes, Mr Holmes, and I should have wormed out the whole story under their very noses. My very next question might have cleared the matter up, but at that instant the door opened and a woman stepped into the room. I could not see her clearly enough to know more than that she was tall and graceful, with black hair, and clad in some sort of loose white gown.

' "Harold!" said she, speaking English with a broken accent, "I could not stay away longer. It is so lonely up there with only—oh, my God, it is Paul!"

'These last words were in Greek, and at the same instant the man, with a convulsive effort, tore the plaster from his lips, and screaming out "Sophy! Sophy!" rushed into the woman's arms. Their embrace was but for an instant, however, for the younger man seized the woman and pushed her out of the room, while the elder easily overpowered his emaciated victim, and dragged him away through the other door. For a moment I was left alone in the room, and I sprang to my feet with some vague idea that I might in some way get a clue to what this house was in which I found myself. Fortunately, however, I took no steps, for, looking up, I saw that the older man was standing in the doorway, with his eyes fixed upon me.

' "That will do, Mr Melas," said he. "You perceive that we have taken you into our confidence over some very private business. We should not have troubled you only that our friend who speaks Greek and who began these negotiations has been forced to return to the East. It was quite necessary for us to find someone to take his place, and we were fortunate in hearing of your powers."

'I bowed.

' "There are five sovereigns here," said he, walking up to me, "which will, I hope, be a sufficient fee. But remember," he added, tapping me lightly on the chest and giggling, "if you speak to a human soul about this—one human soul, mind—well, may God have mercy upon your soul!"

'I cannot tell you the loathing and horror with which this insignificant-looking man inspired me. I could see him better now as the lamp-light shone upon him. His features were peaky and sallow, and his little, pointed beard was thready and ill-nourished. He pushed his face forward as he spoke, and his lips and eyelids were continually twitching, like a man with St Vitus's Dance. I could not help thinking that his strange, catchy little laugh was also a symptom of some nervous malady. The terror of his face lay in his eyes, however, steel grey, and glistening coldly, with a malignant, inexorable cruelty in their depths.

' "We shall know if you speak of this," said he. "We have our own means of information. Now, you will find the carriage waiting, and my friend will see you on your way."

'I was hurried through the hall, and into the vehicle, again obtaining that momentary glimpse of trees and a garden. Mr Latimer followed closely at my heels, and took his place opposite to me without a word. In silence we again drove for an interminable distance, with the windows raised, until at last, just after midnight, the carriage pulled up.

' "You will get down here, Mr Melas," said my companion. "I am sorry to leave you so far from your house, but there is no alternative. Any attempt upon your part to follow the carriage can only end in injury to yourself."

'He opened the door as he spoke, and I had hardly time to spring out when the coachman lashed the horse, and the carriage rattled away. I looked round me in astonishment. I was on some sort of a heathy common, mottled over with dark clumps of furze bushes. Far away stretched a line of houses, with a light here and there in the upper windows. On the other side I saw the red signal lamps of a railway.

'The carriage which had brought me was already out of sight. I stood gazing round and wondering where on earth I might be, when I saw someone coming towards me in the darkness. As he came up to me I made out that it was a railway porter.

' "Can you tell me what place this is?" I asked.

' "Wandsworth Common,*" said he.

' "Can I get a train into town?"

' "If you walk on a mile or so, to Clapham Junction,*" said he, "you'll just be in time for the last to Victoria."

'So that was the end of my adventure, Mr Holmes. I do not know where I was nor whom I spoke with, nor anything, save what I have told you. But I know that there is foul play going on, and I want to help that unhappy man if I can. I told the whole story to Mr Mycroft Holmes next morning, and, subsequently, to the police.'

We all sat in silence for some little time after listening to this extraordinary narrative. Then Sherlock looked across at his brother.

'Any steps?' he asked.

Mycroft picked up the *Daily News*, which was lying on a side table.

' "Anybody supplying any information as to the whereabouts of a Greek gentleman named Paul Kratides, from Athens, who is unable to speak English, will be rewarded. A similar reward paid to anyone giving information about a Greek lady whose first name is Sophy. X2473." That was in all the dailies. No answer.'

'How about the Greek Legation?'

'I have inquired. They know nothing.'

'A wire to the head of the Athens police, then.'

'Sherlock has all the energy of the family,' said Mycroft, turning to me. 'Well, you take up the case by all means, and let me know if you do any good.'

'Certainly,' answered my friend, rising from his chair. 'I'll let you know, and Mr Melas also. In the meantime, Mr Melas, I should certainly be on my guard if I were you, for, of course, they must know through these advertisements that you have betrayed them.'

As we walked home together Holmes stopped at a telegraph office and sent off several wires.

'You see, Watson,' he remarked, 'our evening has been by no means wasted. Some of my most interesting cases have come to me in this way through Mycroft. The problem which we have just listened to, although it can admit of but one explanation, has still some distinguishing features.'

'You have hopes of solving it?'

'Well, knowing as much as we do, it will be singular indeed if we fail to discover the rest. You must yourself have formed some theory which will explain the facts to which we have listened.'

'In a vague way, yes.'

'What was your idea, then?'

'It seemed to me to be obvious that this Greek girl had been carried off by the young Englishman named Harold Latimer.'

'Carried off from where?'

'Athens, perhaps.'

Sherlock Holmes shook his head. 'This young man could not talk a word of Greek. The lady could talk English fairly well. Inference that she had been in England some little time, but he had not been in Greece.'

'Well, then, we will presume that she had come on a visit to England, and that this Harold had persuaded her to fly with him.'

'That is the more probable.'

'Then the brother—for that, I fancy, must be the relationship—comes over from Greece to interfere. He imprudently puts himself into the power of the young man and his older associate. They seize him and use violence towards him in order to make him sign some papers to make over the girl's fortune—of which he may be trustee—to them. This he refuses to do. In order to negotiate with him, they have to get an interpreter, and they pitch upon this Mr Melas, having used some other one before. The girl is not told of the arrival of her brother, and finds it out by the merest accident.'

'Excellent, Watson,' cried Holmes. 'I really fancy that you are not far from the truth. You see that we hold all the cards, and we have only to fear some sudden act of violence on their part. If they give us time we must have them.'

'But how can we find where this house lies?'

'Well, if our conjecture is correct, and the girl's name is, or was, Sophy Kratides, we should have no difficulty in tracing her. That must be our main hope, for the brother, of course, is a complete stranger. It is clear that some time has elapsed since this Harold established these relations with the girl—some weeks at any rate—since the brother in Greece has had time to hear of it, and come across. If they have been living in the same place during this time, it is probable that we shall have some answer to Mycroft's advertisement.'

We had reached our house in Baker Street whilst we had been talking. Holmes ascended the stairs first, and as he opened the door of our room he gave a start of surprise. Looking over his shoulder I was equally astonished. His brother Mycroft was sitting smoking in the arm-chair.

'Come in, Sherlock! Come in, sir,' said he blandly, smiling at our surprised faces. 'You don't expect such energy from me, do you, Sherlock? But somehow this case attracts me.'

'How did you get here?'

'I passed you in a hansom.'

'There has been some new development?'

'I had an answer to my advertisement.'

'Ah!'

'Yes; it came within a few minutes of your leaving.'

'And to what effect?'

Mycroft Holmes took out a sheet of paper.

'Here it is,' said he, 'written with a J pen on royal cream paper, by a middle-aged man with a weak constitution. "Sir," he says, "in answer to your advertisement of today's date, I beg to inform you that I know the young lady in question very well. If you should care to call upon me, I could give you some particulars as to her painful history. She is living at present at The Myrtles, Beckenham.*— Yours faithfully, J. DAVENPORT."

'He writes from Lower Brixton,' said Mycroft Holmes. 'Do you not think that we might drive to him now, Sherlock, and learn these particulars?'

'My dear Mycroft, the brother's life is more valuable than the sister's story. I think we should call at Scotland Yard for Inspector Gregson, and go straight out to Beckenham. We know that a man is being done to death, and every hour may be vital.'

'Better pick up Mr Melas upon our way,' I suggested; 'we may need an interpreter.'

'Excellent!' said Sherlock Holmes. 'Send the boy for a four-wheeler, and we shall be off at once.' He opened the table-drawer as he spoke, and I noticed that he slipped his revolver into his pocket. 'Yes,' said he, in answer to my glance, 'I should say from what we have heard that we are dealing with a particularly dangerous gang.'

It was almost dark before we found ourselves in Pall Mall, at the rooms of Mr Melas. A gentleman had just called for him, and he was gone.

'Can you tell me where?' asked Mycroft Holmes.

'I don't know, sir,' answered the woman who had opened the door. 'I only know that he drove away with the gentleman in a carriage.'

'Did the gentleman give a name?'

'No, sir.'

'He wasn't a tall, handsome, dark young man?'

'Oh, no, sir; he was a little gentleman, with glasses, thin in the face, but very pleasant in his ways, for he was laughing all the time that he was talking.'

'Come along!' cried Sherlock Holmes abruptly. 'This grows serious!' he observed, as we drove to Scotland Yard. 'These men have got hold of Melas again. He is a man of no physical courage, as they are well aware from their experience the other night. This villain was able to terrorize him the instant that he got into his presence. No doubt they want his professional services; but, having used him, they may be inclined to punish him for what they will regard as his treachery.'

Our hope was that by taking train we might get to Beckenham as soon as, or sooner than, the carriage. On reaching Scotland Yard, however, it was more than an hour before we could get Inspector Gregson and comply with the legal formalities which would enable us to enter the house. It was a quarter to ten before we reached London Bridge, and half-past before the four of us alighted on the Beckenham platform. A drive of half a mile brought us to The Myrtles—a large, dark house, standing back from the road in its own grounds. Here we dismissed our cab, and made our way up the drive together.

'The windows are all dark,' remarked the inspector. 'The house seems deserted.'

'Our birds are flown and the nest empty,' said Holmes.

'Why do you say so?'

'A carriage heavily loaded with luggage has passed out during the last hour.'

The inspector laughed. 'I saw the wheel-tracks in the light of the gate-lamp, but where does the luggage come in?'

'You may have observed the same wheel-tracks going the other way. But the outward-bound ones were very much deeper—so much so that we can say for a certainty that there was a very considerable weight on the carriage.'

'You get a trifle beyond me there,' said the inspector, shrugging his shoulders. 'It will not be an easy door to force. But we will try if we cannot make someone hear us.'

He hammered loudly at the knocker and pulled at the bell, but without any success. Holmes had slipped away, but he came back in a few minutes.

'I have a window open,' said he.

'It is a mercy that you are on the side of the force, and not against it, Mr Holmes,' remarked the inspector, as he noted the clever way in which my friend had forced back the catch. 'Well, I think that, under the circumstances, we may enter without waiting for an invitation.'

One after the other we made our way into a large apartment, which was evidently that in which Mr Melas had found himself. The inspector had lit his lantern, and by its light we could see the two doors, the curtain, the lamp and the suit of Japanese mail as he had described them. On the table stood two glasses, an empty brandy bottle, and the remains of a meal.

'What is that?' asked Holmes suddenly.

We all stood still and listened. A low, moaning sound was coming from somewhere above our heads. Holmes rushed to the door and out into the hall. The dismal noise came from upstairs. He dashed up, the inspector and I at his heels, while his brother, Mycroft, followed as quickly as his great bulk would permit.

Three doors faced us upon the second floor, and it was from the central of these that the sinister sounds were issuing, sinking sometimes into a dull mumble and rising again into a shrill whine. It was locked, but the key was on the outside. Holmes flung open the door and rushed in, but he was out again in an instant with his hand to his throat.

'It's charcoal!' he cried. 'Give it time. It will clear.'

Peering in, we could see that the only light in the room came from a dull, blue flame, which flickered from a small brass tripod in the centre. It threw a livid, unnatural circle upon the floor, while in the shadows beyond we saw the vague loom of two figures, which crouched against the wall. From the open door there reeked a horrible, poisonous exhalation, which set us gasping and coughing. Holmes rushed to the top of the stairs to draw in the fresh air, and then, dashing into the room, he threw up the window and hurled the brazen tripod out into the garden.

'We can enter in a minute,' he gasped, darting out again. 'Where is a candle?* I doubt if we could strike a match in that atmosphere. Hold the light at the door and we shall get them out, Mycroft. Now!'

With a rush we got to the poisoned men and dragged them out on to the landing. Both of them were blue-lipped and insensible, with swollen, congested faces and protruding eyes. Indeed, so distorted were their features that, save for his black beard and stout figure, we might have failed to recognize in one of them the Greek interpreter who had parted from us only a few hours before at the Diogenes Club. His hands and feet were securely strapped together, and he bore over one eye the mark of a violent blow. The other, who was secured in a similar fashion, was a tall man in the last stage of emaciation, with several strips of sticking-plaster arranged in a grotesque pattern over his face. He had ceased to moan as we laid him down, and a glance showed me that for him, at least, our aid had come too late. Mr Melas, however, still lived, and in less than an hour, with the aid of ammonia and brandy,* I had the satisfaction of seeing him open his eyes, and of knowing that my hand had drawn him back from the dark valley in which all paths meet.

It was a simple story which he had to tell, and one which did but confirm our own deductions. His visitor on entering his rooms had drawn a life preserver* from his sleeve, and had so impressed him with the fear of instant and inevitable death, that he had kidnapped him for the second time. Indeed, it was almost mesmeric the effect which this giggling ruffian had produced upon the unfortunate linguist, for he could not speak of him save with trembling hands and a blanched cheek. He had been taken swiftly to Beckenham, and had acted as interpreter in a second interview, even more dramatic than the first, in which the two Englishmen had menaced their prisoner with instant death if he did not comply with their demands. Finally, finding him proof against every threat, they had hurled him back into his prison, and after reproaching Melas with his treachery, which appeared from the newspaper advertisements, they had stunned him with a blow from a stick, and he remembered nothing more until he found us bending over him.

And this was the singular case of the Grecian Interpreter, the explanation of which is still involved in some mystery. We were able to find out, by communicating with the gentleman who had answered the advertisement, that the unfortunate young lady came of a wealthy Grecian family, and that she had been on a visit to some friends in England. While there she had met a young man named Harold Latimer, who had acquired an ascendancy over her and had eventually persuaded her to fly with him. Her friends, shocked at the event, had contented themselves with informing her brother at Athens, and had then washed their hands of the matter. The brother, on his arrival in England, had imprudently placed himself in the power of Latimer, and of his associate, whose name was Wilson Kemp—a man of the foulest antecedents. These two, finding that through his ignorance of the language he was helpless in their hands, had kept him a prisoner, and had endeavoured, by cruelty and starvation, to make him sign away his own and his sister's property. They had kept him in the house without the girl's knowledge, and the plaster over the face had been for the purpose of making recognition difficult in case she should ever catch a glimpse of him. Her feminine perceptions, however, had instantly seen through the disguise when, on the occasion of the interpreter's first visit, she had seen him for the first time. The poor girl, however, was herself a prisoner, for there was no one about the house except the man who acted as coachman, and his wife, both of whom were tools of the conspirators. Finding that their secret was out and that their prisoner was not to be coerced, the two villains, with the girl, had fled away at a few hours' notice from the furnished house which they had hired, having first, as they thought, taken vengeance both upon the man who had defied and the one who had betrayed them.

Months afterwards a curious newspaper cutting reached us from Buda-Pesth. It told how two Englishmen who had been travelling with a woman had met with a tragic end. They had each been stabbed, it seems, and the Hungarian police were of opinion that they had quarrelled and had inflicted mortal injuries upon each other. Holmes, however, is, I fancy, of a different way of thinking, and he holds to this day that if one could find the Grecian girl one might learn how the wrongs of herself and her brother came to be avenged.

The Naval Treaty

THE July which immediately succeeded my marriage was made memorable by three cases of interest in which I had the privilege of being associated with Sherlock Holmes and of studying his methods. I find them recorded in my notes under the headings of 'The Adventure of the Second Stain',* 'The Adventure of the Naval Treaty', and 'The Adventure of the Tired Captain'. The first of these, however, deals with interests of such importance, and implicates so many of the first families in the kingdom, that for many years it will be impossible to make it public. No case, however, in which Holmes was ever engaged has illustrated the value of his analytical methods so clearly or has impressed those who were associated with him so deeply. I still retain an almost verbatim report of the interview in which he demonstrated the true facts of the case to Monsieur Dubuque, of the Paris Police, and Fritz von Waldbaum, the well-known specialist of Dantzig, both of whom had wasted their energies upon what proved to be side issues. The new century will have come, however, before the story can be safely told. Meanwhile, I pass on to the second upon my list, which promised also, at one time, to be of national importance, and was marked by several incidents which give it a quite unique character.

During my school days I had been intimately associated with a lad named Percy Phelps, who was of much the same age as myself, though he was two classes ahead of me. He was a very brilliant boy, and carried away every prize which the school had to offer, finishing his exploits by winning a scholarship, which sent him on to continue his triumphant career at Cambridge. He was, I remember, extremely well connected, and even when we were all little boys together, we knew that his mother's brother was Lord Holdhurst,* the great Conservative politician. This gaudy relationship did him little good at school; on the contrary, it seemed rather a piquant thing to us to chevy him about the playground and hit him over the shins with a wicket. But it was another thing when he came out into the world. I heard vaguely that his abilities and the influence which he commanded had won him a good position at the Foreign Office, and then he passed completely out of my mind until the following letter recalled his existence:

'Briarbrae, Woking.*

'MY DEAR WATSON,—I have no doubt that you can remember "Tadpole" Phelps, who was in the fifth form when you were in the third. It is possible even that you may have heard that, through my uncle's influence, I obtained a good appointment at the Foreign Office, and that I was in a situation of trust and honour until a horrible misfortune came suddenly to blast my career.

'There is no use writing the details of that dreadful event. In the event of your acceding to my request, it is probable that I shall have to narrate them to you. I have only just recovered from nine weeks of brain fever, and am still exceedingly weak. Do you think that you could bring your friend, Mr Holmes, down to see me? I should like to have his opinion of the case, though the authorities assure me that nothing more can be done. Do try to bring him down, and as soon as possible. Every minute seems an hour while I live in this horrible suspense. Assure him that, if I have not asked his advice sooner, it was not because I did not appreciate his talents, but because I have been off my head ever since the blow fell. Now I am clear again, though I dare not think of it too much for fear of a relapse. I am still so weak that I have to write, as you see, by dictating. Do try and bring him.

'Your old schoolfellow,
'PERCY PHELPS.'

There was something that touched me as I read this letter, something pitiable in the reiterated appeals to bring Holmes.

So moved was I that, even if it had been a difficult matter, I should have tried it; but, of course, I knew well that Holmes loved his art so, that he was ever as ready to bring his aid as his client could be to receive it. My wife agreed with me that not a moment should be lost in laying the matter before him, and so, within an hour of breakfast-time, I found myself back once more in the old rooms in Baker Street.

Holmes was seated at his side table clad in his dressing-gown and working hard over a chemical investigation. A large curved retort was boiling furiously in the bluish flame of a Bunsen burner, and the distilled drops were condensing into a two-litre measure. My friend hardly glanced up as I entered, and I, seeing that his investigation must be of importance, seated myself in an arm-chair and waited. He dipped into this bottle or that, drawing out a few drops of each with his glass pipette, and finally brought a test-tube containing a solution over to the table. In his right hand he had a slip of litmus-paper.

'You come at a crisis, Watson,' said he. 'If this paper remains blue, all is well. If it turns red, it means a man's life.' He dipped it into the test-tube and it flushed at once into a dull, dirty crimson. 'Hum! I thought as much!' he cried. 'I shall be at your service in one instant, Watson. You will find tobacco in the Persian slipper.' He turned to his desk and scribbled off several telegrams, which were handed over to the page-boy. Then he threw himself down in the chair opposite, and drew up his knees until his fingers clasped round his long, thin shins.

'A very commonplace little murder,' said he. 'You've got something better, I fancy. You are the stormy petrel of crime, Watson. What is it?'

I handed him the letter, which he read with the most concentrated attention.

'It does not tell us very much, does it?' he remarked, as he handed it back to me.

'Hardly anything.'

'And yet the writing is of interest.'

'But the writing is not his own.'

'Precisely. It is a woman's.'

'A man's, surely!' I cried.

'No, a woman's; and a woman of rare character. You see, at the commencement of an investigation, it is something to know that your client is in close contact with someone who for good or evil has an exceptional nature. My interest is already awakened in the case. If you are ready, we will start at once for Woking* and see this diplomatist who is in such evil case, and the lady to whom he dictates his letters.'

We were fortunate enough to catch an early train at Waterloo*, and in a little under an hour we found ourselves among the fir-woods and the heather of Woking. Briarbrae proved to be a large detached house standing in extensive grounds, within a few minutes' walk of the station. On sending in our cards we were shown into an elegantly-appointed drawing-room, where we were joined in a few minutes by a rather stout man, who received us with much hospitality. His age may have been nearer forty than thirty, but his cheeks were so ruddy and his eyes so merry, that he still conveyed the impression of a plump and mischievous boy.

'I am so glad that you have come,' said he, shaking our hands with effusion. 'Percy has been inquiring for you all the morning. Ah, poor old chap, he clings to any straw. His father and mother asked me to see you, for the mere mention of the subject is very painful to them.'

'We have had no details yet,' observed Holmes. 'I perceive that you are not yourself a member of the family.'

Our acquaintance looked surprised, and then glancing down he began to laugh.

'Of course you saw the "J.H." monogram on my locket,' said he. 'For a moment I thought you had done something clever. Joseph Harrison is my name, and as Percy is to marry my sister Annie, I shall at least be a relation by marriage. You will find my sister in his room, for she has nursed him hand-and-foot these two months back.* Perhaps we had better go in at once, for I know how impatient he is.'

The chamber into which we were shown was on the same floor as the drawing-room. It was furnished partly as a sitting and partly as a bedroom, with flowers arranged daintily in every nook and corner. A young man, very pale and worn, was lying upon a sofa near the open window, through which came the rich scent of the garden and the balmy summer air. A woman was sitting beside him, and rose as we entered.

'Shall I leave, Percy?' she asked.

He clutched her hand to detain her. 'How are you, Watson?' said he cordially. 'I should never have known you under that moustache, and I dare say you would not be prepared to swear to me. This, I presume, is your celebrated friend, Mr Sherlock Holmes?'

I introduced him in a few words, and we both sat down. The stout young man had left us, but his sister still remained, with her hand in that of the invalid. She was a striking-looking woman, a little short and thick for symmetry, but with a beautiful olive complexion, large, dark Italian eyes, and a wealth of deep black hair. Her rich tints made the white face of her companion the more worn and haggard by the contrast.

'I won't waste your time,' said he, raising himself upon the sofa. 'I'll plunge into the matter without further preamble. I was a happy and successful man, Mr Holmes, and on the eve of being married, when a sudden and dreadful misfortune wrecked all my prospects in life.

'I was, as Watson may have told you, in the Foreign Office, and through the influence of my uncle, Lord Holdhurst, I rose rapidly to a responsible position. When my uncle became Foreign Minister in this Administration he gave me several missions of trust, and as I always brought them to a successful conclusion, he came at last to have the utmost confidence in my ability and tact.

'Nearly ten weeks ago—to be more accurate, on the 23rd of May—he called me into his private room and, after complimenting me upon the good work which I had done, informed me that he had a new commission of trust for me to execute.

' "This," said he, taking a grey roll of paper from his bureau, "is the original of that secret treaty between England and Italy,* of which, I regret to say, some rumours have already got into the public Press. It is of enormous importance that nothing further should leak out. The French or Russian Embassies would pay an immense sum to learn the contents of these papers. They should not leave my bureau were it not that it is absolutely necessary to have them copied. You have a desk in your office?"

' "Yes, sir."

' "Then take the treaty and lock it up there. I shall give directions that you may remain behind when the others go, so that you may copy it at your leisure, without fear of being overlooked. When you have finished, re-lock both the original and the draft in the desk, and hand them over to me personally tomorrow morning."

'I took the papers and—'

'Excuse me an instant,' said Holmes; 'were you alone during this conversation?'

'Absolutely.'

'In a large room?'

'Thirty feet each way.'

'In the centre?'

'Yes, about it.'

'And speaking low?'

'My uncle's voice is always remarkably low. I hardly spoke at all.'

'Thank you,' said Holmes, shutting his eyes; 'pray go on.'

'I did exactly what he had indicated, and waited until the other clerks had departed. One of them in my room, Charles Gorot,* had some arrears of work to make up, so I left him there and went out to dine. When I returned he was gone. I was anxious to hurry my work, for I knew that Joseph, the Mr Harrison whom you saw just now, was in town, and that he would travel down to Woking by the eleven o'clock train, and I wanted if possible to catch it.

'When I came to examine the treaty I saw at once that it was of such importance that my uncle had been guilty of no exaggeration in what he had said. Without going into details, I may say that it defined the position of Great Britain towards the Triple Alliance,* and foreshadowed the policy which this country would pursue in the event of the French fleet gaining a complete ascendancy over that of Italy in the Mediterranean. The questions treated in it were purely naval. At the end were the signatures of the high dignitaries who had signed it. I glanced my eyes over it, and then settled down to my task of copying.

'It was a long document, written in the French language,* and containing twenty-six separate articles. I copied as quickly as I could, but at nine o'clock I had only done nine articles, and it seemed hopeless for me to attempt to catch my train. I was feeling drowsy and stupid, partly from my dinner and also from the effects of a long day's work. A cup of coffee would clear my brain. A commissionaire remains all night in a little lodge at the foot of the stairs, and is in the habit of making coffee at his spirit-lamp for any of the officials who may be working overtime. I rang the bell, therefore, to summon him.

'To my surprise, it was a woman who answered the summons, a large, coarse-faced, elderly woman, in an apron. She explained that she was the commissionaire's wife, who did the charing, and I gave her the order for the coffee.

'I wrote two more articles, and then, feeling more drowsy than ever, I rose and walked up and down the room to stretch my legs. My coffee had not yet come, and I wondered what the cause of the delay could be. Opening the door, I started down the corridor to find out. There was a straight passage dimly lit* which led from the room in which I had been working, and was the only exit from it. It ended in a curving staircase, with the commissionaire's lodge in the passage at the bottom. Half-way down this staircase is a small landing, with another passage running into it at right angles. The second one leads, by means of a second small stair, to a side door used by servants, and also as a short cut by clerks when coming from Charles Street.

'Here is a rough chart of the place.'

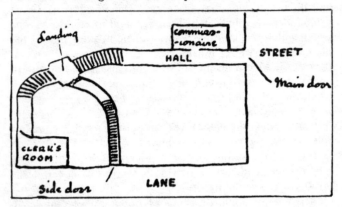

'Thank you. I think that I quite follow you,' said Sherlock Holmes.

'It is of the utmost importance that you should notice this point. I went down the stairs and into the hall, where I found the commissionaire fast asleep in his box, with the kettle boiling furiously upon the spirit lamp, for the water was spurting over the floor. I had put out my hand* and was about to shake the man, who was still sleeping soundly, when a bell over his head rang loudly, and he woke with a start.

' "Mr Phelps, sir!" said he, looking at me in bewilderment.

' "I came down to see if my coffee was ready."

' "I was boiling the kettle when I fell asleep, sir," He looked at me and then up at the still quivering bell, with an ever-growing astonishment upon his face.

' "If you was here, sir, then who rang the bell?" he asked.

' "The bell!" I said. "What bell is it?"

' "It's the bell of the room you were working in."

'A cold hand seemed to close round my heart. Someone, then, was in that room where my precious treaty lay upon the table. I ran frantically up the stairs and along the passage. There was no one in the corridor, Mr Holmes. There was no one in the room. All was exactly as I left it, save only that the papers committed to my care had been taken from the desk on which they lay. The copy was there and the original was gone.'

Holmes sat up in his chair and rubbed his hands. I could see that the problem was entirely to his heart. 'Pray, what did you do then?' he murmured.

'I recognized in an instant that the thief must have come up the stairs from the side door. Of course, I must have met him if he had come the other way.'

'You were satisfied that he could not have been concealed in the room all the time, or in the corridor which you have just described as dimly lighted?'

'It is absolutely impossible. A rat could not conceal himself either in the room or the corridor. There is no cover at all.'

'Thank you. Pray proceed.'

'The commissionaire, seeing by my pale face that something was to be feared, had followed me upstairs. Now we both rushed along the corridor and down the steep steps which led to Charles Street. The door at the bottom was closed but unlocked. We flung it open and rushed out. I can distinctly remember that as we did so there came three chimes from a neighbouring church. It was a quarter to ten.'

'That is of enormous importance,' said Holmes, making a note upon his shirt cuff.

'The night was very dark, and a thin, warm rain was falling. There was no one in Charles Street, but a great traffic was going on, as usual, in Whitehall at the extremity. We rushed along the pavement, bareheaded as we were, and at the far corner we found a policeman standing.

' "A robbery has been committed," I gasped. "A document of immense value has been stolen from the Foreign Office. Has anyone passed this way?"

' "I have been standing here for a quarter of an hour, sir," said he; "only one person has passed during that time—a woman, tall and elderly, with a Paisley shawl."

' "Ah, that is only my wife," cried the commissionaire. "Has no one else passed?"

' "No one."

' "Then it must be the other way that the thief took," cried the fellow, tugging at my sleeve.

'But I was not satisfied, and the attempts which he made to draw me away increased my suspicions.

' "Which way did the woman go?" I cried.

' "I don't know, sir. I noticed her pass, but I had no special reason for watching her. She seemed to be in a hurry."

' "How long ago was it?"

' "Oh, not very many minutes."

' "Within the last five?"

' "Well, it could not be more than five."

' "You're only wasting your time, sir, and every minute now is of importance," cried the commissionaire. "Take my word for it that my old woman has nothing to do with it, and come down to the other end of the street. Well, if you won't, I will," and with that he rushed off in the other direction.

'But I was after him in an instant and caught him by the sleeve.

' "Where do you live?" said I.

' "No. 16, Ivy Lane, Brixton," he answered; "but don't let yourself be drawn away upn a false scent, Mr Phelps. Come to the other end of the street, and let us see if we can hear of anything."

'Nothing was to be lost by following his advice. With the policeman we both hurried down, but only to find the street full of traffic, many people coming and going, but all only too eager to get to a place of safety upon so wet a night. There was no lounger who could tell us who had passed.

'Then we returned to the office, and searched the stairs and the passage without result. The corridor which led to the room was laid down with a kind of creamy linoleum which shows an impression very easily. We examined it very carefully, but found no outline of any footmark.'

'Had it been raining all the evening?'

'Since about seven.'

'How is it, then, that the woman who came into the room about nine left no traces with her muddy boots?'

'I am glad you raise the point. It occurred to me at the time. The charwomen are in the habit of taking off their boots at the commissionaire's office, and putting on list slippers.'*

'That is very clear. There were no marks, then, though the night was a wet one? The chain of events is certainly one of extraordinary interest. What did you do next?'

'We examined the room also. There was no possibility of a secret door, and the windows are quite thirty feet from the ground. Both of them were fastened on the inside. The carpet prevents any possibility of a trap-door, and the ceiling is of the ordinary white-washed kind. I will pledge my life that whoever stole my papers could only have come through the door.'

'How about the fireplace?'

'They use none. There is a stove. The bell-rope hangs from the wire just to the right of my desk. Whoever rang it must have come right up to the desk to do it. But why should any criminal wish to ring the bell? It is a most insoluble mystery.'

'Certainly the incident was unusual. What were your next steps? You examined the room, I presume, to see if the intruder had left any traces—any cigar end, or dropped glove, or hairpin, or other trifle?'

'There was nothing of the sort.'

'No smell?'

'Well, we never thought of that.'

'Ah, a scent of tobacco would have been worth a great deal to us in such an investigation.'

'I never smoke myself, so I think I should have observed it if there had been any smell of tobacco. There was absolutely no clue of any kind. The only tangible fact was that the commissionaire's wife—Mrs Tangey was the name—had hurried out of the place. He could give no explanation save that it was about the time when the woman always went home. The policeman and I agreed that our best plan would be to seize the woman before she could get rid of the papers, presuming that she had them.

'The alarm had reached Scotland Yard by this time, and Mr Forbes, the detective, came round at once and took up the case with a great deal of energy. We hired a hansom, and in half an hour we were at the address which had been given to us. A young woman opened the door, who proved to be Mrs Tangey's eldest daughter. Her mother had not come back yet, and we were shown into the front room to wait.

'About ten minutes later a knock came at the door, and here we made the one serious mistake for which I blame myself. Instead of opening the door ourselves we allowed the girl to do so. We heard her say, "Mother, there are two men in the house waiting to see you," and an instant afterwards we heard the patter of feet rushing down the passage. Forbes flung open the door, and we both ran into the back room or kitchen, but the woman had got there before us. She stared at us with defiant eyes, and then suddenly recognising me, an expression of absolute astonishment came over her face.

' "Why, if it isn't Mr Phelps, of the office!" she cried.

' "Come, come, who did you think we were when you ran away from us?" asked my companion.

' "I thought you were the brokers,*" said she. "We've had some trouble with a tradesman."

' "That's not quite good enough," answered Forbes. "We have reason to believe that you have taken a paper of importance from the Foreign Office, and that you ran in here to dispose of it. You must come back with us to Scotland Yard to be searched."

'It was in vain that she protested and resisted. A four-wheeler was brought, and we all three drove back in it. We had first made an examination of the kitchen, and especially of the kitchen fire, to see whether she might have made away with the papers during the instant that she was alone.

There were no signs, however, of any ashes or scraps. When we reached Scotland Yard she was handed over at once to the female searcher. I waited in an agony of suspense until she came back with her report. There were no signs of the papers.

'Then, for the first time, the horror of my situation came in its full force upon me. Hitherto I had been acting, and action had numbed thought. I had been so confident of regaining the treaty at once that I had not dared to think of what would be the consequence if I failed to do so. But now there was nothing more to be done, and I had leisure to realize my position. It was horrible! Watson there would tell you that I was a nervous, sensitive boy at school. It is my nature. I thought of my uncle and of his colleagues in the Cabinet, of the shame which I had brought upon him, upon myself, upon everyone connected with me. What though I was the victim of an extraordinary accident? No allowance is made for accidents where diplomatic interests are at stake. I was ruined; shamefully, hopelessly ruined. I don't know what I did. I fancy I must have made a scene. I have a dim recollection of a group of officials who crowded round me endeavouring to soothe me. One of them drove down with me to Waterloo and saw me into the Woking train. I believe that he would have come all the way had it not been that Dr Ferrier, who lives near me, was going down by that very train. The doctor most kindly took charge of me, and it was well he did so, for I had a fit in the station, and before we reached home I was practically a raving maniac.

'You can imagine the state of things here when they were roused from their beds by the doctor's ringing, and found me in this condition. Poor Annie here and my mother were broken-hearted. Dr Ferrier had just heard enough from the detective at the station to be able to give an idea of what had happened, and his story did not mend matters. It was evident to all that I was in for a long illness, so Joseph was bundled out of this cheery bedroom, and it was turned into a sickroom for me. Here I have lain, Mr Holmes, for over nine weeks, unconscious, and raving with brain fever. If it had not been for Miss Harrison here and for the doctor's care I should not be speaking to you now. She has nursed me by day, and a hired nurse has looked after me by night, for in my mad fits I was capable of anything. Slowly my reason has cleared, but it is only during the last three days that my memory has quite returned. Sometimes I wish that it never had. The first thing I did was to wire to Mr Forbes, who had the case in hand. He came out and assured me that though everything has been done, no trace of a clue has been discovered. The commissionaire and his wife have been examined in every way without any light being thrown upon the matter. The suspicions of the police then rested upon young Gorot, who, as you may remember, stayed overtime in the office that night. His remaining behind and his French name were really the only two points which could suggest suspicion; but as a matter of fact, I did not begin work until he had gone, and his people are of Huguenot extraction,* but as English in sympathy and tradition as you and I are. Nothing was found to implicate him in any way, and there the matter dropped. I turn to you, Mr Holmes, as absolutely my last hope. If you fail me, then my honour as well as my position are for ever forfeited.'

The invalid sank back upon his cushions, tired out by this long recital, while his nurse poured him out a glass of some stimulating medicine. Holmes sat silently with his head thrown back and his eyes closed in an attitude which might seem listless to a stranger, but which I knew betokened the most intense absorption.

'Your statement has been so explicit,' said he at last, 'that you have really left me very few questions to ask. There is one of the very utmost importance, however. Did you tell anyone that you had this special task to perform?'

'No one.'

'Not Miss Harrison here, for example?'

'No. I had not been back to Woking between getting the order and executing the commission.'

'And none of your people had by chance been to see you?'

'None.'

'Did any of them know their way about in the office?'

'Oh, yes; all of them had been shown over it.'

'Still, of course, if you said nothing to anyone about the treaty, these inquiries are irrelevant.'

'I said nothing.'

'Do you know anything of the commissionaire?'

'Nothing, except that he is an old soldier.'

'What regiment?'

'Oh, I have heard—Coldstream Guards.'*

'Thank you. I have no doubt I can get details from Forbes. The authorities are excellent at amassing facts, though they do not always use them to advantage. What a lovely thing a rose is!'

He walked past the couch to the open window, and held up the drooping stalk of a moss rose, looking down at the dainty blend of crimson and green. It was a new phase of his character to me, for I had never before seen him show any keen interest in natural objects.

'There is nothing in which deduction is so necessary as in religion,' said he, leaning with his back against the shutters. 'It can be built up as an exact science by the reasoner. Our highest assurance of the goodness of Providence seems to me to rest in the flowers. All other things, our powers, our desires, our food, are really necessary for our existence in the first instance. But this rose is an extra. Its smell and its colour are an embellishment of life, not a condition of it. It is only goodness which gives extras, and so I say again that we have much to hope from the flowers.'

Percy Phelps and his nurse looked at Holmes during this demonstration with surprise and a good deal of disappointment written upon their faces. He had fallen into a reverie, with the moss rose between his fingers. It had lasted some minutes before the young lady broke in upon it.

'Do you see any prospect of solving this mystery, Mr Holmes?' she asked, with a touch of asperity in her voice.

'Oh, the mystery!' he answered, coming back with a start to the realities of life. 'Well, it would be absurd to deny that the case is a very abstruse and complicated one; but I can promise you that I will look into the matter and let you know any points which may strike me.'

'Do you see any clue?'

'You have furnished me with seven, but of course I must test them before I can pronounce upon their value.'

'You suspect someone?'

'I suspect myself——'

'What?'

'Of coming to conclusions too rapidly.'

'Then go to London and test your conclusions.'

'Your advice is very excellent, Miss Harrison,' said Holmes, rising. 'I think, Watson, we cannot do better. Do not allow yourself to indulge in false hopes, Mr Phelps. The affair is a very tangled one.'

'I shall be in a fever until I see you again,' cried the diplomatist.

'Well, I'll come out by the same train to-morrow, though it's more than likely that my report will be a negative one.'

'God bless you for promising to come,' cried our client. 'It gives me fresh life to know that something is being done. By the way, I have had a letter from Lord Holdhurst.'

'Ha! What did he say?'

'He was cold, but not harsh. I dare say my severe illness prevented him from being that. He repeated that the matter was of the utmost importance, and added that no steps would be taken about my future—by which he means, of course, my dismissal—until my health was restored and I had an opportunity of repairing my misfortune.'

'Well, that was reasonable and considerate,' said Holmes. 'Come, Watson, for we have a good day's work before us in town.'

Mr Joseph Harrison drove us down to the station, and we were soon whirling up in a Portsmouth train. Holmes was sunk in profound thought, and hardly opened his mouth until we had passed Clapham Junction.

'It's a very cheering thing to come into London by any of these lines which run high and allow you to look down upon the houses like this.'

I thought he was joking, for the view was sordid enough, but he soon explained himself.

'Look at those big, isolated clumps of buildings rising up above the slates, like brick islands in a lead-coloured sea.'

'The Board schools.'*

'Lighthouses, my boy! Beacons of the future! Capsules, with hundreds of bright little seeds in each, out of which will spring the wiser, better England of the future. I suppose that man Phelps does not drink?'

'I should not think so.'

'Nor should I. But we are bound to take every possibility into account. The poor devil has certainly got himself into very deep water, and it's a question whether we shall ever be able to get him ashore. What did you think of Miss Harrison?'

'A girl of strong character.'

'Yes, but she is a good sort, or I am mistaken. She and her brother are the only children of an ironmaster somewhere up Northumberland way. Phelps* got engaged to her when travelling last winter, and she came down to be introduced to his people, with her brother as escort. Then came the smash, and she stayed on to nurse her lover, while brother Joseph, finding himself pretty snug, stayed on too. I've been making a few independent inquiries, you see. But today must be a day of inquiries.'

'My practice—' I began.

'Oh, if you find your own cases more interesting than mine—' said Holmes, with some asperity.

'I was going to say that my practice could get along very well for a day or two, since it is the slackest time in the year.'

'Excellent,' said he, recovering his good humour. 'Then we'll look into this matter together. I think that we should begin by seeing Forbes. He can probably tell us all the details we want, until we know from what side the case is to be approached.'

'You said you had a clue.'

'Well, we have several, but we can only test their value by further inquiry. The most difficult crime to track is the one which is purposeless. Now, this is not purposeless. Who is it that profits by it? There is the French Ambassador, there is the Russian, there is whoever might sell it to either of these, and there is Lord Holdhurst.'

'Lord Holdhurst!'

'Well, it is just conceivable that a statesman might find himself in a position where he was not sorry to have such a document accidentally destroyed.'

'Not a statesman with the honourable record of Lord Holdhurst.'

'It is a possibility, and we cannot afford to disregard it. We shall see the noble lord to-day, and find out if he can tell us anything. Meanwhile, I have already set inquiries upon foot.'

'Already?'

'Yes, I sent wires from Woking Station to every evening paper in London. This advertisement will appear in each of them.'

He handed over a sheet torn from the note-book. On it was scribbled in pencil:

'£10 Reward.—The number of the cab which dropped a fare at or about the door of the Foreign Office in Charles Street, at a quarter to ten in the evening of May 23rd. Apply 221B, Baker Street.'

'You are confident that the thief came in a cab?'

'If not, there is no harm done. But if Mr Phelps is correct in stating that there is no hiding-place either in the room or the corridors, then the person must have come from outside. If he came from outside on so wet a night, and yet left no trace of damp upon the linoleum, which was examined within a few minutes of his passing, then it is exceedingly probable that he came in a cab. Yes, I think that we may safely deduce a cab.'

'It sounds plausible.'

'That is one of the clues of which I spoke. It may lead us to something. And then of course there is the bell—which is the most distinctive feature of the case. Why should the bell ring? Was it the thief that did it out of bravado? Or was it someone who was with the thief who did it in order to prevent the crime? Or was it an accident? Or was it—?' He sank back into the state of intense and silent thought from which he had emerged, but it seemed to me, accustomed as I was to his every mood, that some new possibility had dawned suddenly upon him.

It was twenty past three when we reached our terminus, and after a hasty luncheon at the buffet we pushed on at once to Scotland Yard. Holmes had already wired to Forbes, and we found him waiting to receive us: a small, foxy man, with a sharp but by no means amiable expression. He was decidedly frigid in his manner to us, especially when he heard the errand upon which we had come.

'I've heard of your methods before now, Mr Holmes,' said he tartly. 'You are ready enough to use all the information that the police can lay at your disposal, and then you try to finish the case yourself and bring discredit upon them.'

'On the contrary,' said Holmes; 'out of my last fifty-three cases my name has only appeared in four, and the police have had all the credit in forty-nine. I don't blame you for not knowing this, for you are young and inexperienced; but if you wish to get on in your new duties you will work with me, and not against me.'

'I'd be very glad of a hint or two,' said the detective, changing his manner. 'I've certainly had no credit from the case so far.'

'What steps have you taken?'

'Tangey, the commissionaire, has been shadowed. He left the Guards with a good character, and we can find nothing against him. His wife is a bad lot, though. I fancy she knows more about this than appears.'

'Have you shadowed her?'

'We have set one of our women on to her. Mrs Tangey drinks, and our woman has been with her twice when she was well on, but she could get nothing out of her.'

'I understand that they have had brokers in the house?'

'Yes, but they were paid off.'

'Where did the money come from?'

'That was all right. His pension was due; they have not shown any sign of being in funds.'

'What explanation did she give of having answered the bell when Mr Phelps rang for the coffee?'

'She said that her husband was very tired and she wished to relieve him.'

'Well, certainly that would agree with his being found, a little later, asleep in his chair. There is nothing against them, then, but the woman's character. Did you ask her why she hurried away that night? Her haste attracted the attention of the police-constable.'

'She was later than usual, and wanted to get home.'

'Did you point out to her that you and Mr Phelps, who started at least twenty minutes after her, got home before her?'

'She explains that by the difference between a 'bus and a hansom.'

'Did she make it clear why, on reaching her house, she ran into the back kitchen?'

'Because she had the money there with which to pay off the brokers.'

'She has at least an answer for everything. Did you ask her whether in leaving she met anyone or saw anyone loitering about Charles Street?'

'She saw no one but the constable.'

'Well, you seem to have cross-examined her pretty thoroughly. What else have you done?'

'The clerk, Gorot, has been shadowed all these nine weeks, but without result. We can show nothing against him.'

'Anything else?'

'Well, we have nothing else to go upon—no evidence of any kind.'

'Have you formed any theory about how that bell rang?'

'Well, I must confess that it beats me. It was a cool hand, whoever it was, to go and give the alarm like that.'

'Yes, it was a queer thing to do. Many thanks to you for what you have told me. If I can put the man into your hands you shall hear from me. Come along, Watson!'

'Where are we going to now?' I asked, as we left the office.

'We are now going to interview Lord Holdhurst, the Cabinet Minister and future Premier of England.'

We were fortunate in finding that Lord Holdhurst was still in his chambers at Downing Street,* and on Holmes, sending in his card we were instantly shown up. The statesman received us with that old-fashioned courtesy for which he is remarkable, and seated us on the two luxurious easy chairs on either side of the fireplace. Standing on the rug between us, with his slight, tall figure, his sharp-featured, thoughtful face, and his curling hair prematurely tinged with grey, he seemed to represent that not too common type, a nobleman who is in truth noble.

'Your name is very familiar to me, Mr Holmes,' said he, smiling. 'And, of course, I cannot pretend to be ignorant of the object of your visit. There has only been one occurrence in these offices which could call for your attention. In whose interest are you acting, may I ask?'

'In that of Mr Percy Phelps,' answered Holmes.

'Ah, my unfortunate nephew! You can understand that our kinship makes it the more impossible for me to screen him in any way. I fear that the incident must have a very prejudicial effect upon his career.'

'But if the document is found?'

'Ah, that, of course, would be different.'

'I had one or two questions which I wished to ask you, Lord Holdhurst.'

'I shall be happy to give you any information in my power.'

'Was it in this room that you gave your instructions as to the copying of the document?'

'It was.'

'Then you could have hardly been overheard?'

'It is out of the question.'

'Did you ever mention to anyone that it was your intention to give out the treaty to be copied?'

'Never.'

'You are certain of that?'

'Absolutely.'

'Well, since you never said so, and Mr Phelps never said so, and nobody else knew anything of the matter, then the thief's presence in the room was purely accidental. He saw his chance and he took it.'

The statesman smiled. 'You take me out of my province there,' said he.

Holmes considered for a moment. 'There is another very important point which I wish to discuss with you,' said he. 'You feared, as I understand, that very grave results might follow from the details of this treaty becoming known?'

A shadow passed over the expressive face of the statesman. 'Very grave results, indeed.'

'And have they occurred?'

'Not yet.'

'If the treaty had reached, let us say, the French or Russian Foreign Office, you would expect to hear of it?'

'I should,' said Lord Holdhurst, with a wry face.

'Since nearly ten weeks have elapsed, then, and nothing has been heard, it is not unfair to suppose that for some reason the treaty has not reached them?'

Lord Holdhurst shrugged his shoulders.

'We can hardly suppose, Mr Holmes, that the thief took the treaty in order to frame it and hang it up.'

'Perhaps he is waiting for a better price.'

'If he waits a little longer he will get no price at all. The treaty will cease to be a secret in a few months.'

'That is most important,' said Holmes. 'Of course it is a possible supposition that the thief has had a sudden illness—'

'An attack of brain fever, for example?' asked the statesman, flashing a swift glance at him.

'I did not say so,' said Holmes imperturbably. 'And now, Lord Holdhurst, we have already taken up too much of your valuable time, and we shall wish you good-day.'

'Every success to your investigation, be the criminal who it may,' answered the nobleman, as he bowed us out at the door.

'He's a fine fellow,' said Holmes, as we came out into Whitehall. 'But he has a struggle to keep up his position. He is far from rich, and has many calls. You noticed, of course, that his boots had been re-soled? Now, Watson, I won't detain you from your legitimate work any longer. I shall do nothing more to-day, unless I have an answer to my cab advertisement. But I should be extremely obliged to you if you would come down with me to Woking to-morrow, by the same train which we took to-day.'

I met him accordingly next morning, and we travelled down to Woking together. He had had no answer to his advertisement, he said, and no fresh light had been thrown upon the case. He had, when he so willed it, the utter immobility of countenance of a Red Indian, and I could not gather from his appearance whether he was satisfied or not with the position of the case. His conversation, I remember, was about the Bertillon system of measurements,* and he expressed his enthusiastic admiration of the French savant.

We found our client still under the charge of his devoted nurse, but looking considerably better than before. He rose from the sofa and greeted us without difficulty when we entered.

'Any news?' he asked eagerly.

'My report, as I expected, is a negative one,' said Holmes. 'I have seen Forbes, and I have seen your uncle, and I have set one or two trains of inquiry upon foot which may lead to something.'

'You have not lost heart, then?'

'By no means.'

'God bless you for saying that!' cried Miss Harrison. 'If we keep our courage and our patience, the truth must come out.'

'We have more to tell you than you have for us,' said Phelps, reseating himself upon the couch.

'I hoped you might have something.'

'Yes, we have had an adventure during the night, and one which might have proved to be a serious one.' His expression grew very grave as he spoke, and a look of something akin to fear sprang up in his eyes. 'Do you know,' said he, 'that I begin to believe that I am the unconscious centre of some monstrous conspiracy, and that my life is aimed at as well as my honour?'

'Ah!' cried Holmes.

'It sounds incredible, for I have not, as far as I know, an enemy in the world. Yet from last night's experience I can come to no other conclusion.'

'Pray let me hear it.'

'You must know that last night was the very first night that I have ever slept without a nurse in the room. I was so much better that I thought I could dispense with one. I had a night-light burning, however. Well, about two in the morning I had sunk into a light sleep, when I was suddenly aroused by a slight noise. It was like the sound which a mouse makes when it is gnawing a plank, and I lay listening to it for some time under the impression that it must come from that cause. Then it grew louder, and suddenly there came from the window a sharp metallic snick. I sat up in amazement. There could be no doubt what the sounds were now. The faint ones had been caused by someone forcing an instrument through the slit between the sashes, and the second by the catch being pressed back.

'There was a pause then for about ten minutes, as if the person were waiting to see whether the noise had awoken me. Then I heard a gentle creaking as the window was very slowly opened. I could stand it no longer, for my nerves are not what they used to be. I sprang out of bed and flung open the shutters. A man was crouching at the window. I could see little of him, for he was gone like a flash. He was wrapped in some sort of cloak, which came across the lower part of his face. One thing only I am sure of, and that is that he had some weapon in his hand. It looked to me like a long knife. I distinctly saw the gleam of it as he turned to run.'

'This is most interesting,' said Holmes. 'Pray, what did you do then?'

'I should have followed him through the open window if I had been stronger. As it was, I rang the bell and roused the house. It took me some little time, for the bell rings in the kitchen, and the servants all sleep upstairs. I shouted, however, and that brought Joseph down, and he roused the others. Joseph and the groom found marks on the flower-bed outside the window, but the weather has been so dry lately that they found it hopeless to follow the trail across the grass. There's a place, however, on the wooden fence which skirts the road which shows signs, they tell me, as if someone had got over and had snapped the top of the rail in doing so. I have said nothing to the local police yet, for I thought I had best have your opinion first.'

This tale of our client's appeared to have an extraordinary effect upon Sherlock Holmes. He rose from his chair and paced about the room in incontrollable excitement.

'Misfortunes never come singly,' said Phelps, smiling, though it was evident that his adventure had somewhat shaken him.

'You have certainly had your share,' said Holmes. 'Do you think you could walk round the house with me?'

'Oh, yes, I should like a little sunshine. Joseph will come too.'

'And I also,' said Miss Harrison.

'I am afraid not,' said Holmes, shaking his head. 'I think I must ask you to remain sitting exactly where you are.'

The young lady resumed her seat with an air of displeasure. Her brother, however, had joined us, and we set off all four together. We passed round the lawn to the outside of the young diplomatist's window. There were, as he had said, marks upon the flower-bed, but they were hopelessly blurred and vague. Holmes stooped over them for an instant, and then rose, shrugging his shoulders.

'I don't think anyone could make much of this,' said he. 'Let us go round the house and see why this particular room was chosen by the burglar. I should have thought those larger windows of the drawing-room and dining-room would have had more attractions for him.'

'They are more visible from the road,' suggested Mr Joseph Harrison.

'Ah, yes, of course. There is a door here which he might have attempted. What is it for?'

'It is the side entrance for tradespeople. Of course, it is locked at night.'

'Have you ever had an alarm like this before?'

'Never,' said our client.

'Do you keep plate in the house, or anything to attract burglars?'

'Nothing of value.'

Holmes strolled round the house with his hands in his pockets, and a negligent air which was unusual with him.

'By the way,' said he to Joseph Harrison, 'you found some place, I understand, where the fellow scaled the fence. Let us have a look at that.'

The plump young man led us to a spot where the top of one of the wooden rails had been cracked. A small fragment of the wood was hanging down. Holmes pulled it off and examined it critically.

'Do you think that was done last night? It looks rather old, does it not?'

'Well, possibly so.'

'There are no marks of anyone jumping down upon the other side. No, I fancy we shall get no help here. Let us go back to the bedroom and talk the matter over.'

Percy Phelps was walking very slowly, leaning upon the arm of his future brother-in-law. Holmes walked swiftly across the lawn, and we were at the open window of the bedroom long before the others came up.

'Miss Harrison,' said Holmes, speaking with the utmost intensity of manner, 'you must stay where you are all day. Let nothing prevent you from staying where you are all day. It is of most vital importance.'

'Certainly, if you wish it, Mr Holmes,' said the girl in astonishment.

'When you go to bed lock the door of this room on the outside and keep the key. Promise to do this.'

'But Percy?'

'He will come to London with us.'

'And I am to remain here?'

'It is for his sake. You can serve him! Quick! Promise!'

She gave a nod of assent just as the other two came up.

'Why do you sit moping there, Annie?' cried her brother. 'Come out into the sunshine!'

'No, thank you, Joseph. I have a slight headache, and this room is deliciously cool and soothing.'

'What do you propose now, Mr Holmes?' asked our client.

'Well, in investigating this minor affair we must not lose sight of our main inquiry. It would be a very great help to me if you would come to London with us.'

'At once?'

'Well, as soon as you conveniently can. Say in an hour.'

'I feel quite strong enough, if I can really be of any help.'

'The greatest possible.'

'Perhaps you would like me to stay there to-night.'

'I was just going to propose it.'

'Then if my friend of the night comes to revisit me, he will find the bird flown. We are all in your hands, Mr Holmes, and you must tell us exactly what you would like done. Perhaps you would prefer that Joseph came with us, so as to look after me?'

'Oh, no; my friend Watson is a medical man, you know, and he'll look after you. We'll have our lunch here, if you will permit us, and then we shall all three set off for town together.'

It was arranged as he suggested, though Miss Harrison excused herself from leaving the bedroom, in accordance with Holmes's suggestion. What the object of my friend's manœuvres was I could not conceive, unless it were to keep the lady away from Phelps, who, rejoiced by his returning health and by the prospect of action, lunched with us in the dining-room. Holmes had a still more startling surprise for us, however, for after accompanying us down to the station and seeing us into our carriage, he calmly announced that he had no intention of leaving Woking.

'There are one or two small points which I should desire to clear up before I go,' said he. 'Your absence, Mr Phelps,

will in some ways rather assist me. Watson, when you reach
London you would oblige me by driving at once to Baker
Street with our friend here, and remaining with him until I
see you again. It is fortunate that you are old schoolfellows,
as you must have much to talk over. Mr Phelps can have the
spare bedroom to-night, and I shall be with you in time for
breakfast, for there is a train which will take me into
Waterloo at eight.'

'But how about our investigation in London?' asked
Phelps, ruefully.

'We can do that to-morrow. I think that just at present I
can be of more immediate use here.'

'You might tell them at Briarbrae that I hope to be back
to-morrow night,' cried Phelps, as we began to move from
the platform.

'I hardly expect to go back to Briarbrae,' answered
Holmes, and waved his hand to us cheerily as we shot out
from the station.

Phelps and I talked it over on our journey, but neither of us
could devise a satisfactory reason for this new development.

'I suppose he wants to find out some clue as to the
burglary last night, if a burglar it was. For myself, I don't
believe it was an ordinary thief.'

'What is your idea, then?'

'Upon my word, you may put it down to my weak nerves
or not, but I believe there is some deep political intrigue
going on around me, and that, for some reason that passes
my understanding, my life is aimed at by the conspirators.
It sounds high-flown and absurd, but consider the facts!
Why should a thief try to break in at a bedroom window,
where there could be no hope of any plunder, and why
should he come with a long knife in his hand?'

'You are sure it was not a housebreaker's jemmy?'

'Oh, no; it was a knife. I saw the flash of the blade quite
distinctly.'

'But why on earth should you be pursued with such
animosity?'

'Ah! that is the question.'

'Well, if Holmes takes the same view, that would account for his action, would it not? Presuming that your theory is correct, if he can lay his hands upon the man who threatened you last night, he will have gone a long way towards finding who took the naval treaty. It is absurd to suppose that you have two enemies, one of whom robs you while the other threatens your life.'

'But Mr Holmes said that he was not going to Briarbrae.'

'I have known him for some time,' said I, 'but I never knew him do anything yet without a very good reason,' and with that our conversation drifted off into other topics.

But it was a weary day for me. Phelps was still weak after his long illness, and his misfortunes made him querulous and nervous. In vain I endeavoured to interest him in Afghanistan, in India, in social questions, in anything which might take his mind out of the groove. He would always come back to his lost treaty; wondering, guessing, speculating, as to what Holmes was doing, what steps Lord Holdhurst was taking, what news we should have in the morning. As the evening wore on his excitement became quite painful.

'You have implicit faith in Holmes?' he asked.

'I have seen him do some remarkable things.'

'But he never brought light into anything quite so dark as this?'

'Oh, yes; I have known him solve questions which presented fewer clues than yours.'

'But not where such large interests are at stake?'

'I don't know that. To my certain knowledge he has acted on behalf of three of the reigning Houses of Europe* in very vital matters.'

'But you know him well, Watson. He is such an inscrutable fellow, that I never quite know what to make of him. Do you think he is hopeful? Do you think he expects to make a success of it?'

'He has said nothing.'

'That is a bad sign.'

'On the contrary, I have noticed that when he is off the trail he generally says so. It is when he is on a scent, and is not quite absolutely sure yet that it is the right one, that he is most taciturn. Now, my dear fellow, we can't help matters by making ourselves nervous about them, so let me implore you to go to bed, and so be fresh for whatever may await us to-morrow.'

I was able at last to persuade my companion to take my advice, though I knew from his excited manner that there was not much hope of sleep for him. Indeed, his mood was infectious, for I lay tossing half the night myself, brooding over this strange problem, and inventing a hundred theories, each of which was more impossible than the last. Why had Holmes remained at Woking? Why had he asked Miss Harrison to stay in the sick room all day? Why had he been so careful not to inform the people at Briarbrae that he intended to remain near them? I cudgelled my brains until I fell asleep in the endeavour to find some explanation which would cover all these facts.

It was seven o'clock when I awoke, and I set off at once for Phelps's room, to find him haggard and spent after a sleepless night. His first question was whether Holmes had arrived yet.

'He'll be here when he promised,' said I, 'and not an instant sooner or later.'

And my words were true, for shortly after eight a hansom dashed up to the door and our friend got out of it. Standing in the window, we saw that his left hand was swathed in a bandage and that his face was very grim and pale. He entered the house, but it was some little time before he came upstairs.

'He looks like a beaten man,' cried Phelps.

I was forced to confess that he was right. 'After all,' said I, 'the clue of the matter lies probably here in town.'

Phelps gave a groan.

'I don't know how it is,' said he, 'but I had hoped for so much from his return. But surely his hand was not tied up like that yesterday? What can be the matter?'

'You are not wounded, Holmes?' I asked, as my friend entered the room.

'Tut, it is only a scratch through my own clumsiness,' he answered, nodding his good morning to us. 'This case of yours, Mr Phelps, is certainly one of the darkest which I have ever investigated.'

'I feared that you would find it beyond you.'

'It has been a most remarkable experience.'

'That bandage tells of adventures,' said I. 'Won't you tell us what has happened?'

'After breakfast, my dear Watson. Remember that I have breathed thirty miles of Surrey air this morning. I suppose there has been no answer to my cabman advertisement? Well, well, we cannot expect to score every time.'

The table was all laid, and, just as I was about to ring, Mrs Hudson entered with the tea and coffee. A few minutes later she brought in the covers, and we all drew up to the table, Holmes ravenous, I curious, and Phelps in the gloomiest state of depression.

'Mrs Hudson has risen to the occasion,' said Holmes, uncovering a dish of curried chicken. 'Her cuisine is a little limited, but she has as good an idea of breakfast as a Scotch-woman.* What have you there, Watson?'

'Ham and eggs,' I answered.

'Good! What are you going to take, Mr Phelps: curried fowl, eggs, or will you help yourself?'

'Thank you, I can eat nothing,' said Phelps.

'Oh, come! Try the dish before you.'

'Thank you, I would really rather not.'

'Well, then,' said Holmes, with a mischievous twinkle, 'I suppose that you have no objection to helping me?'

Phelps raised the cover, and as he did so he uttered a scream, and sat there staring with a face as white as the plate upon which he looked. Across the centre of it was lying a little cylinder of blue-grey paper. He caught it up, devoured it with his eyes, and then danced madly about the room, pressing it to his bosom and shrieking out in his delight. Then he fell back into an arm-chair, so limp and exhausted with his own emotions that we had to pour brandy down his throat to keep him from fainting.

'There! there!' said Holmes, soothingly, patting him upon the shoulder. 'It was too bad to spring it on you like this; but Watson here will tell you that I never can resist a touch of the dramatic.'

Phelps seized his hand and kissed it. 'God bless you!' he cried; 'you have saved my honour.'

'Well, my own was at stake, you know,' said Holmes. 'I assure you, it is just as hateful to me to fail in a case as it can be to you to blunder over a commission.'

Phelps thrust away the precious document into the innermost pocket of his coat.

'I have not the heart to interrupt your breakfast any further, and yet I am dying to know how you got it and where it was.'

Sherlock Holmes swallowed a cup of coffee and turned his attention to the ham and eggs. Then he rose, lit his pipe, and settled himself down into his chair.

'I'll tell you what I did first, and how I came to do it afterwards,' said he. 'After leaving you at the station I went for a charming walk through some admirable Surrey scenery to a pretty little village called Ripley,* where I had my tea at an inn, and took the precaution of filling my flask and of putting a paper of sandwiches in my pocket. There I remained until evening, when I set off for Woking again and found myself in the high road outside Briarbrae just after sunset.

'Well, I waited until the road was clear—it is never a very frequented one at any time, I fancy—and then I clambered over the fence into the grounds.'

'Surely the gate was open?' ejaculated Phelps.

'Yes; but I have a peculiar taste in these matters. I chose the place where the three fir trees stand, and behind their screen I got over without the least chance of anyone in the house being able to see me. I crouched down among the bushes on the other side, and crawled from one to the other— witness the disreputable state of my trouser knees— until I had reached the clump of rhododendrons just opposite to your bedroom window. There I squatted down and awaited developments.

'The blind was not down in your room, and I could see Miss Harrison sitting there reading by the table. It was a quarter past ten when she closed her book, fastened the shutters, and retired. I heard her shut the door, and felt quite sure that she had turned the key in the lock.'

'The key?' ejaculated Phelps.

'Yes, I had given Miss Harrison instructions to lock the door on the outside and take the key with her when she went to bed. She carried out every one of my injunctions to the letter, and certainly without her co-operation you would not have that paper in your coat pocket. She departed then, the lights went out, and I was left squatting in the rhododendron bush.

'The night was fine, but still it was a very weary vigil. Of course, it has the sort of excitement about it that the sportsman feels when he lies beside the watercourse and waits for the big game. It was very long, though—almost as long, Watson, as when you and I waited in that deadly room when we looked into the little problem of the "Speckled Band".* There was a church clock down at Woking which struck the quarters, and I thought more than once that it had stopped. At last, however, about two in the morning, I suddenly heard the gentle sound of a bolt being pushed back, and the creaking of a key. A moment later the servants' door was opened and Mr Joseph Harrison stepped out into the moonlight.'

'Joseph!' ejaculated Phelps.

'He was bare-headed, but he had a black cloak thrown over his shoulder, so that he could conceal his face in an instant if there were any alarm. He walked on tip-toe under the shadow of the wall, and when he reached the window he worked a long-bladed knife through the sash and pushed back the catch. Then he flung open the window and, putting his knife through the crack in the shutters, he thrust the bar up and swung them open.

'From where I lay I had a perfect view of the inside of the room and of every one of his movements. He lit the two candles which stand upon the mantelpiece, and then he proceeded to turn back the corner of the carpet in the neighbourhood of the door. Presently he stooped and picked out a square piece of board, such as is usually left to enable plumbers to get at the joints of the gas pipes. This one covered, as a matter of fact, the T-joint* which gives off the pipe which supplies the kitchen underneath. Out of this hiding-place he drew that little cylinder of paper, pushed down the board, rearranged the carpet, blew out the candles, and walked straight into my arms as I stood waiting for him outside the window.

'Well, he has rather more viciousness than I gave him credit for, has Master Joseph. He flew at me with his knife, and I had to grass him twice,* and got a cut over the knuckles, before I had the upper hand of him. He looked "murder" out of the only eye he could see with when we had finished, but he listened to reason and gave up the papers. Having got them I let my man go, but I wired full particulars to Forbes this morning. If he is quick enough to catch his bird, well and good! But if, as I shrewdly suspect, he finds the nest empty before he gets there, why, all the better for the Government. I fancy that Lord Holdhurst, for one, and Mr Percy Phelps, for another, would very much rather that the affair never got so far as a police-court.'

'My God!' gasped our client. 'Do you tell me that during these long ten weeks of agony the stolen papers were within the very room with me all the time?'

'So it was.'

'And Joseph! Joseph a villain and a thief!'

'Hum! I am afraid Joseph's character is a rather deeper and more dangerous one than one might judge from his appearance. From what I have heard from him this morning, I gather that he has lost heavily in dabbling with stocks, and that he is ready to do anything on earth to better his fortunes. Being an absolutely selfish man, when a chance presented itself he did not allow either his sister's happiness or your reputation to hold his hand.'

Percy Phelps sank back in his chair. 'My head whirls,' said he; 'your words have dazed me.'

'The principal difficulty in your case,' remarked Holmes, in his didactic fashion, 'lay in the fact of there being too much evidence. What was vital was overlaid and hidden by what was irrelevant. Of all the facts which were presented to us, we had to pick just those which we deemed to be essential, and then piece them together in their order, so as to reconstruct this very remarkable chain of events. I had already begun to suspect Joseph, from the fact that you had intended to travel home with him that night, and that therefore it was a likely enough thing that he should call for you—knowing the Foreign Office well—upon his way. When I heard that someone had been so anxious to get into the bedroom, in which no one but Joseph could have concealed anything—you told us in your narrative how you had turned Joseph out when you arrived with the doctor—my suspicions all changed to certainties, especially as the attempt was made on the first night upon which the nurse was absent, showing that the intruder was well acquainted with the ways of the house.'

'How blind I have been!'

'The facts of the case, as far as I have worked them out, are these: This Joseph Harrison entered the office through the Charles Street door, and knowing his way he walked straight into your room the instant after you left it. Finding no one there he promptly rang the bell, and at the instant that he did so his eyes caught the paper upon the table. A glance showed him that chance had put in his way a State document of immense value, and in a flash* he had thrust it into his pocket and was gone. A few minutes elapsed, as you remember, before the sleepy commissionaire drew your attention to the bell, and those were just enough to give the thief time to make his escape.

'He made his way to Woking by the first train, and, having examined his booty and assured himself that it really was of immense value, he concealed it in what he thought was a very safe place, with the intention of taking it out again in a day or two, and carrying it to the French Embassy, or wherever he thought that a long price was to be had. Then came your sudden return. He, without a moment's warning, was bundled out of his room, and from that time onwards there were always at least two of you there to prevent him from regaining his treasure. The situation to him must have been a maddening one. But at last he thought he saw his chance. He tried to steal in, but was baffled by your wakefulness. You may remember that you did not take your usual draught that night.'

'I remember.'

'I fancy that he had taken steps to make that draught efficacious, and that he quite relied upon your being unconscious. Of course, I understood that he would repeat the attempt whenever it could be done with safety. Your leaving the room gave him the chance he wanted. I kept Miss Harrison in it all day, so that he might not anticipate us. Then, having given him the idea that the coast was clear, I kept guard as I have described. I already knew that the papers were probably in the room, but I had no desire to rip up all the planking and skirting in search of them. I let him take them, therefore, from the hiding-place, and so saved myself an infinity of trouble. Is there any other point which I can make clear?'

'Why did he try the window on the first occasion,' I asked, 'when he might have entered by the door?'

'In reaching the door he would have to pass seven bedrooms. On the other hand, he could get out on to the lawn with ease. Anything else?'

'You do not think,' asked Phelps, 'that he had any murderous intention? The knife was only meant as a tool.'

'It may be so,' answered Holmes, shrugging his shoulders. 'I can only say for certain that Mr Joseph Harrison is a gentleman to whose mercy I should be extremely unwilling to trust.'

The Final Problem

IT is with a heavy heart that I take up my pen to write these the last words in which I shall ever record the singular gifts by which my friend Mr Sherlock Holmes was distinguished. In an incoherent and, as I deeply feel, an entirely inadequate fashion, I have endeavoured to give some account of my strange experiences in his company from the chance which first brought us together at the period of the 'Study in Scarlet', up to the time of his interference in the matter of the 'Naval Treaty'—an interference which had the unquestionable effect of preventing a serious international complication. It was my intention to have stopped there, and to have said nothing of that event which has created a void in my life which the lapse of two years has done little to fill. My hand has been forced, however, by the recent letters in which Colonel James Moriarty defends the memory of his brother, and I have no choice but to lay the facts before the public exactly as they occurred. I alone know the absolute truth of the matter, and I am satisfied that the time has come when no good purpose is to be served by its suppression. As far as I know, there have been only three accounts in the public Press: that in the *Journal de Genève** upon May 6th, 1891, the Reuter's* despatch in the English papers upon May 7th, and finally the recent letters to which I have alluded. Of these the first and second were extremely condensed, while the last is, as I shall now show, an absolute perversion of the facts. It lies with me to tell for the first time what really took place between Professor Moriarty and Mr Sherlock Holmes.

It may be remembered that after my marriage, and my subsequent start in private practice, the very intimate relations which had existed between Holmes and myself became to some extent modified. He still came to me from time to time when he desired a companion in his investigations, but these occasions grew more and more seldom, until I find that in the year 1890 there were only three cases of which I retain any record. During the winter of that year and the early spring of 1891, I saw in the papers that he had been engaged by the French Government upon a matter of supreme importance, and I received two notes from Holmes, dated from Narbonne and from Nîmes,* from which I gathered that his stay in France was likely to be a long one. It was with some surprise, therefore, that I saw him walk into my consulting-room upon the evening of the 24th of April. It struck me that he was looking even paler and thinner than usual.

'Yes, I have been using myself up rather too freely,' he remarked, in answer to my look rather than to my words; 'I have been a little pressed of late. Have you any objection to my closing your shutters?'

The only light in the room came from the lamp upon the table at which I had been reading. Holmes edged his way round the wall, and flinging the shutters together, he bolted them securely.

'You are afraid of something?' I asked.

'Well, I am.'

'Of what?'

'Of air-guns.'*

'My dear Holmes, what do you mean?'

'I think that you know me well enough, Watson, to understand that I am by no means a nervous man. At the same time, it is stupidity rather than courage to refuse to recognise danger when it is close upon you. Might I trouble you for a match?' He drew in the smoke of his cigarette as if the soothing influence was grateful to him.

'I must apologize for calling so late,' said he, 'and I must further beg you to be so unconventional as to allow me to leave your house presently by scrambling over your back garden wall.'

'But what does it all mean?' I asked.

He held out his hand, and I saw in the light of the lamp that two of his knuckles were burst and bleeding.

'It's not an airy nothing, you see,' said he, smiling. 'On the contrary, it is solid enough for a man to break his hand over. Is Mrs Watson in?'

'She is away upon a visit.'

'Indeed! You are alone?'

'Quite.'

'Then it makes it the easier for me to propose that you should come away with me for a week on to the Continent.'

'Where?'

'Oh, anywhere. It's all the same to me.'

There was something very strange in all this. It was not Holmes's nature to take an aimless holiday, and something about his pale, worn face told me that his nerves were at their highest tension. He saw the question in my eyes, and, putting his finger-tips together and his elbows upon his knees, he explained the situation.

'You have probably never heard of Professor Moriarty?' said he.

'Never.'

'Aye, there's the genius and the wonder of the thing!' he cried. 'The man pervades London, and no one has heard of him. That's what puts him on a pinnacle in the records of crime. I tell you, Watson, in all seriousness, that if I could beat that man, if I could free society of him, I should feel that my own career had reached its summit, and I should be prepared to turn to some more placid line in life. Between ourselves, the recent cases in which I have been of assistance to the Royal Family of Scandinavia,* and to the French Republic, have left me in such a position that I could continue to live in the quiet fashion which is most congenial to me, and to concentrate my attention upon my chemical researches. But I could not rest, Watson, I could not sit quiet in my chair, if I thought that such a man as Professor Moriarty were walking the streets of London unchallenged.'

'What has he done, then?'

'His career has been an extraordinary one. He is a man of good birth and excellent education, endowed by Nature with a phenomenal mathematical faculty. At the age of twenty-one he wrote a treatise upon the Binomial Theorem,* which has had a European vogue. On the strength of it, he won the Mathematical Chair at one of our smaller Universities, and had, to all appearance, a most brilliant career before him. But the man had hereditary tendencies of the most diabolical kind. A criminal strain ran in his blood, which, instead of being modified, was increased and rendered infinitely more dangerous by his extraordinary mental powers. Dark rumours gathered round him in the University town, and eventually he was compelled to resign his Chair and to come down to London, where he set up as an army coach. So much is known to the world, but what I am telling you now is what I have myself discovered.

'As you are aware, Watson, there is no one who knows the higher criminal world of London so well as I do. For years past I have continually been conscious of some power behind the malefactor, some deep organizing power which for ever stands in the way of the law, and throws its shield over the wrong-doer. Again and again in cases of the most varying sorts—forgery cases, robberies, murders—I have felt the presence of this force, and I have deduced its action in many of those undiscovered crimes in which I have not been personally consulted. For years I have endeavoured to break through the veil which shrouded it, and at last the time came when I seized my thread and followed it, until it led me, after a thousand cunning windings, to ex-Professor Moriarty of mathematical celebrity.

'He is the Napoleon of crime, Watson. He is the organizer of half that is evil and of nearly all that is undetected in this great city. He is a genius, a philosopher, an abstract thinker. He has a brain of the first order. He sits motionless, like a spider in the centre of its web, but that web has a thousand radiations, and he knows well every quiver of each of them. He does little himself. He only plans. But his agents are numerous and splendidly organized. Is there a crime to be done, a paper to be abstracted, we will say, a house to be rifled, a man to be removed—the word is passed to the

Professor, the matter is organized and carried out. The agent may be caught. In that case money is found for his bail or his defence. But the central power which uses the agent is never caught—never so much as suspected.* This was the organization which I deduced, Watson, and which I devoted my whole energy to exposing and breaking up.

'But the Professor was fenced round with safeguards so cunningly devised that, do what I would, it seemed impossible to get evidence which could convict in a court of law. You know my powers, my dear Watson, and yet at the end of three months I was forced to confess that I had at last met an antagonist who was my intellectual equal. My horror at his crimes was lost in my admiration at his skill. But at last he made a trip—only a little, little trip—but it was more than he could afford, when I was so close upon him. I had my chance, and, starting from that point, I have woven my net round him until now it is all ready to close. In three days, that is to say on Monday next, matters will be ripe, and the Professor, with all the principal members of his gang, will be in the hands of the police. Then will come the greatest criminal trial of the century, the clearing up of over forty mysteries, and the rope for all of them—but if we move at all prematurely, you understand, they may slip out of our hands even at the last moment.

'Now, if I could have done this without the knowledge of Professor Moriarty, all would have been well. But he was too wily for that. He saw every step which I took to draw my toils round him. Again and again he strove to break away, but I as often headed him off. I tell you, my friend, that if a detailed account of that silent contest could be written, it would take its place as the most brilliant bit of thrust-and-parry work in the history of detection. Never have I risen to such a height, and never have I been so hard pressed by an opponent. He cut deep, and yet I just undercut him. This morning the last steps were taken, and three days only were wanted to complete the business. I was sitting in my room thinking the matter over, when the door opened and Professor Moriarty stood before me.

'My nerves are fairly proof,* Watson, but I must confess to a start when I saw the very man who had been so much in my thoughts standing there on my threshold. His appearance was quite familiar to me. He is extremely tall and thin, his forehead domes out in a white curve, and his two eyes are deeply sunken in his head. He is clean-shaven, pale, and ascetic-looking, retaining something of the professor in his features. His shoulders are rounded from much study, and his face protrudes forward, and is for ever slowly oscillating from side to side in a curiously reptilian fashion. He peered at me with great curiosity in his puckered eyes.

' "You have less frontal development* than I should have expected," said he at last. "It is a dangerous habit to finger loaded firearms in the pocket of one's dressing-gown."

'The fact is that upon his entrance I had instantly recognised the extreme personal danger in which I lay. The only conceivable escape for him lay in silencing my tongue. In an instant I had slipped the revolver from the drawer into my pocket, and was covering him through the cloth. At his remark I drew the weapon out and laid it cocked upon the table. He still smiled and blinked but there was something about his eyes which made me feel very glad that I had it there.

' "You evidently don't know me," said he.

' "On the contrary," I answered, "I think it is fairly evident that I do. Pray take a chair. I can spare you five minutes if you have anything to say."

' "All that I have to say has already crossed your mind," said he.

' "Then possibly my answer has crossed yours," I replied.

' "You stand fast?"

' "Absolutely."

'He clapped his hand into his pocket, and I raised the pistol from the table. But he merely drew out a memorandum-book in which he had scribbled some dates.

' "You crossed my path on the 4th of January," said he. "On the 23rd you incommoded me; by the middle of February I was seriously inconvenienced by you; at the end of March I was absolutely hampered in my plans; and now, at the close of April, I find myself placed in such a position through your continual persecution that I am in positive danger of losing my liberty.* The situation is becoming an impossible one."

' "Have you any suggestion to make?" I asked.

' "You must drop it, Mr Holmes," said he, swaying his face about. "You really must, you know."

' "After Monday," said I.

' "Tut, tut!" said he. "I am quite sure that a man of your intelligence will see that there can be but one outcome to this affair. It is necessary that you should withdraw. You have worked things in such a fashion that we have only one resource left. It has been an intellectual treat to me to see the way in which you have grappled with this affair, and I say, unaffectedly, that it would be a grief to me to be forced to take any extreme measure. You smile, sir, but I assure you that it really would."

' "Danger is part of my trade," I remarked.

' "This is not danger," said he. "It is inevitable destruction. You stand in the way not merely of an individual, but of a mighty organization, the full extent of which you, with all your cleverness, have been unable to realize. You must stand clear, Mr Holmes, or be trodden under foot."

' "I am afraid," said I, rising, "that in the pleasure of this conversation I am neglecting business of importance which awaits me elsewhere."*

'He rose also and looked at me in silence, shaking his head sadly.

' "Well, well," said he at last. "It seems a pity, but I have done what I could. I know every move of your game. You can do nothing before Monday. It has been a duel between you and me, Mr Holmes. You hope to place me in the dock. I tell you that I will never stand in the dock. You hope to beat me. I tell you that you will never beat me. If you are clever enough to bring destruction upon me, rest assured that I shall do as much to you."

' "You have paid me several compliments, Mr Moriarty," said I. "Let me pay you one in return when I say that if I were assured of the former eventuality I would, in the interests of the public, cheerfully accept the latter."

' "I can promise you the one but not the other," he snarled, and so turned his rounded back upon me and went peering and blinking out of the room.

'That was my singular interview with Professor Moriarty. I confess that it left an unpleasant effect upon my mind. His soft, precise fashion of speech leaves a conviction of sincerity which a mere bully could not produce. Of course, you will say: "Why not take police precautions against him?" The reason is that I am well convinced that it is from his agents the blow would fall. I have the best of proofs that it would be so.'

'You have already been assaulted?'

'My dear Watson, Professor Moriarty is not a man who lets the grass grow under his feet. I went out about midday to transact some business in Oxford Street. As I passed the corner which leads from Bentinck Street on to the Welbeck Street crossing, a two-horse van furiously driven whizzed round and was on me like a flash. I sprang for the footpath and saved myself by the fraction of a second. The van dashed round by Marylebone Lane and was gone in an instant. I kept to the pavement after that, Watson, but as I walked down Vere Street a brick came down from the roof of one of the houses and was shattered to fragments at my feet. I called the police and had the place examined. There were slates and bricks piled upon the roof preparatory to some repairs, and they would have me believe that the wind had toppled over one of these. Of course I knew better, but I could prove nothing. I took a cab after that and reached my brother's rooms in Pall Mall, where I spent the day. Now I have come round to you, and on my way I was attacked by a rough with a bludgeon. I knocked him down, and the police have him in custody; but I can tell you with the most absolute confidence that no possible connection will ever be traced between the gentleman upon whose front teeth I have barked my knuckles and the retiring mathematical coach, who is, I dare say, working out problems upon a blackboard ten miles away. You will not wonder, Watson, that my first act on entering your rooms was to close your shutters, and that I have been compelled to ask your permission to leave the house by some less conspicuous exit than the front door.'

I had often admired my friend's courage, but never more than now, as he sat quietly checking off a series of incidents which must have combined to make up a day of horror.

'You will spend the night here?' I said.

'No, my friend, you might find me a dangerous guest. I have my plans laid, and all will be well. Matters have gone so far now that they can move without my help as far as the arrest goes, though my presence is necessary for a conviction. It is obvious, therefore, than I cannot do better than get away for the few days which remain before the police are at liberty to act. It would be a great pleasure to me, therefore, if you could come on to the Continent with me.'

'The practice is quiet,' said I, 'and I have an accommodating neighbour. I should be glad to come.'

'And to start to-morrow morning?'

'If necessary.'

'Oh, yes, it is most necessary. Then these are your instructions, and I beg, my dear Watson, that you will obey them to the letter, for you are now playing a double-handed game with me against the cleverest rogue and the most powerful syndicate of criminals in Europe. Now listen! You will dispatch whatever luggage you intend to take by a trusty messenger unaddressed to Victoria to-night. In the morning you will send for a hansom, desiring your man to take neither the first nor the second which may present itself. Into this hansom you will jump, and you will drive to the Strand end of the Lowther Arcade,* handing the address to the cabman upon a slip of paper, with a request that he will not throw it away. Have your fare ready, and the instant that your cab stops, dash through the Arcade, timing yourself to reach the other side at a quarter-past nine. You will find a small brougham waiting close to the kerb, driven by a fellow with a heavy black cloak tipped at the collar with red. Into this you will step, and you will reach Victoria in time for the Continental express.'

'Where shall I meet you?'

'At the station. The second first-class carriage from the front will be reserved for us.'

'The carriage is our rendezvous, then?'

'Yes.'

It was in vain that I asked Holmes to remain for the evening. It was evident to me that he thought he might bring trouble to the roof he was under, and that that was the motive which impelled him to go. With a few hurried words as to our plans for the morrow he rose and came out with me into the garden, clambering over the wall which leads into Mortimer Street, and immediately whistling for a hansom, in which I heard him drive away.

In the morning I obeyed Holmes's injunctions to the letter. A hansom was procured with such precautions as would prevent its being one which was placed ready for us, and I drove immediately after breakfast to the Lowther Arcade, through which I hurried at the top of my speed. A brougham was waiting with a very massive driver wrapped in a dark cloak, who, the instant that I had stepped in, whipped up the horse and rattled off to Victoria Station. On my alighting there he turned the carriage, and dashed away again without so much as a look in my direction.

So far all had gone admirably. My luggage was waiting for me, and I had no difficulty in finding the carriage which Holmes had indicated, the less so as it was the only one in the train which was marked 'Engaged'. My only source of anxiety now was the non-appearance of Holmes. The station clock marked only seven minutes from the time when we were due to start. In vain I searched among the groups of travellers and leave-takers for the lithe figure of my friend. There was no sign of him. I spent a few minutes in assisting a venerable Italian priest, who was endeavouring to make a porter understand, in his broken English, that his luggage was to be booked through to Paris. Then, having taken another look round, I returned to my carriage, where

I found that the porter, in spite of the ticket, had given me my decrepit Italian friend as a travelling companion. It was useless for me to explain to him that his presence was an intrusion, for my Italian was even more limited than his English, so I shrugged my shoulders resignedly and continued to look out anxiously for my friend. A chill of fear had come over me, as I thought that his absence might mean that some blow had fallen during the night. Already the doors had all been shut and the whistle blown, when—

'My dear Watson,' said a voice, 'you have not even condescended to say good morning.'

I turned in incontrollable astonishment. The aged ecclesiastic had turned his face towards me. For an instant the wrinkles were smoothed away, the nose drew away from the chin, the lower lip ceased to protrude and the mouth to mumble, the dull eyes regained their fire, the drooping figure expanded. The next the whole frame collapsed again, and Holmes had gone as quickly as he had come.

'Good heavens!' I cried. 'How you startled me!'

'Every precaution is still necessary,' he whispered. 'I have reason to think that they are hot upon our trail. Ah, there is Moriarty himself.'

The train had already begun to move as Holmes spoke. Glancing back I saw a tall man pushing his way furiously through the crowd and waving his hand as if he desired to have the train stopped. It was too late, however, for we were rapidly gathering momentum, and an instant later had shot clear of the station.

'With all our precautions, you see that we have cut it rather fine,' said Holmes, laughing. He rose, and throwing off the black cassock* and hat which had formed his disguise, he packed them away in a hand-bag.

'Have you seen the morning paper, Watson?'

'No.'

'You haven't seen about Baker Street, then?'

'Baker Street?'

'They set fire to our rooms last night. No great harm was done.'

259

'Good heavens, Holmes! This is intolerable.'

'They must have lost my track completely after their bludgeon-man was arrested. Otherwise they could not have imagined that I had returned to my rooms. They have evidently taken the precaution of watching you, however, and that is what has brought Moriarty to Victoria. You could not have made any slip in coming?'

'I did exactly what you advised.'

'Did you find your brougham?'

'Yes, it was waiting.'

'Did you recognize your coachman?'

'No.'

'It was my brother Mycroft. It is an advantage to get about in such a case without taking a mercenary into your confidence. But we must plan what we are to do about Moriarty now.'

'As this is an express, and as the boat runs in connection with it, I should think we have shaken him off very effectively.'

'My dear Watson, you evidently did not realize my meaning when I said that this man may be taken as being quite on the same intellectual plane as myself. You do not imagine that if I were the pursuer I should allow myself to be baffled by so slight an obstacle. Why, then, should you think so meanly of him?'

'What will he do?'

'What I should do.'

'What would you do, then?'

'Engage a special.'*

'But it must be late.'

'By no means. This train stops at Canterbury; and there is always at least a quarter of an hour's delay at the boat. He will catch us there.'

'One would think that we were the criminals. Let us have him arrested on his arrival.'

'It would be to ruin the work of three months. We should get the big fish, but the smaller would dart right and left out of the net. On Monday we should have them all. No, an arrest is inadmissible.'

'What then?'

'We shall get out at Canterbury.'

'And then?'

'Well, then we must make a cross-country journey to Newhaven, and so over to Dieppe. Moriarty will again do what I should do. He will get on to Paris, mark down our luggage, and wait for two days at the depôt. In the meantime we shall treat ourselves to a couple of carpet bags, encourage the manufactures of the countries through which we travel, and make our way at our leisure into Switzerland, via Luxembourg and Basle.'

I am too old a traveller to allow myself to be seriously inconvenienced by the loss of my luggage, but I confess that I was annoyed at the idea of being forced to dodge and hide before a man whose record was black with unutterable infamies. It was evident, however, that Holmes understood the situation more clearly than I did.* At Canterbury, therefore, we alighted, only to find that we should have to wait an hour before we could get a train to Newhaven.

I was still looking rather ruefully after the rapidly disappearing luggage van which contained my wardrobe, when Holmes pulled my sleeve and pointed up the line.

'Already, you see,' said he.

Far away from among the Kentish woods there arose* a thin spray of smoke. A minute later a carriage and engine could be seen flying along the open curve which leads to the station. We had hardly time to take our places* behind a pile of luggage when it passed with a rattle and a roar, beating a blast of hot air into our faces.

'There he goes,' said Holmes, as we watched the carriage swing and rock over the points. 'There are limits, you see, to our friend's intelligence. It would have been a *coup-de-maître** had he deduced what I would deduce and acted accordingly.'

'And what would he have done had he overtaken us?'

'There cannot be the least doubt that he would have made a murderous attack upon me. It is, however, a game at which two may play. The question now is whether we should take a premature lunch here, or run our chance of starving before we reach the buffet at Newhaven.'

We made our way to Brussels that night and spent two days there, moving on upon the third day as far as Strasburg.* On the Monday morning Holmes had telegraphed to the London police, and in the evening we found a reply waiting for us at our hotel. Holmes tore it open, and then with a bitter curse hurled it into the grate.

'I might have known it,' he groaned. 'He has escaped!'

'Moriarty!'

'They have secured the whole gang with the exception of him. He has given them the slip. Of course, when I had left the country there was no one to cope with him. But I did think that I had put the game in their hands. I think that you had better return to England, Watson.'

'Why?'

'Because you will find me a dangerous companion now. This man's occupation is gone. He is lost if he returns to London. If I read his character right he will devote his whole energies to revenging himself upon me. He said as much in our short interview, and I fancy that he meant it. I should certainly recommend you to return to your practice.'

It was hardly an appeal to be successful with one who was an old campaigner as well as an old friend. We sat in the Strasburg *salle-à-manger** arguing the question for half an hour, but the same night we had resumed our journey and were well on our way to Geneva.

For a charming week we wandered up the Valley of the Rhône, and then, branching off at Leuk,* we made our way over the Gemmi Pass, still deep in snow, and so, by way of Interlaken, to Meiringen. It was a lovely trip, the dainty green of the spring below, the virgin white of the winter above; but it was clear to me that never for one instant did Holmes forget the shadow which lay across him. In the homely Alpine villages or in the lonely mountain passes, I could still tell, by his quick glancing eyes and his sharp scrutiny of every face that passed us, that he was well convinced that, walk where we would, we could not walk our-selves clear of the danger which was dogging our footsteps.

Once, I remember, as we passed over the Gemmi, and walked along the border of the melancholy Daubensee, a large rock which had been dislodged from the ridge upon our right clattered down and roared into the lake behind us. In an instant Holmes had raced up on to the ridge, and, standing upon a lofty pinnacle, craned his neck in every direction. It was in vain that our guide assured him that a fall of stones was a common chance in the spring-time at that spot. He said nothing, but he smiled at me with the air of a man who sees the fulfilment of that which he had expected.

And yet for all his watchfulness he was never depressed. On the contrary, I can never recollect having seen him in such exuberant spirits. Again and again he recurred to the fact that if he could be assured that society was freed from Professor Moriarty, he would cheerfully bring his own career to a conclusion.

'I think that I may go so far as to say, Watson, that I have not lived wholly in vain,' he remarked. 'If my record were closed to-night I could still survey it with equanimity. The air of London is the sweeter for my presence. In over a thousand cases I am not aware that I have ever used my powers upon the wrong side. Of late I have been tempted to look into the problems furnished by Nature rather than those more superficial ones for which our artificial state of society is responsible. Your memoirs will draw to an end, Watson, upon the day that I crown my career by the capture or extinction of the most dangerous and capable criminal in Europe.'

I shall be brief, and yet exact, in the little which remains for me to tell. It is not a subject on which I would willingly dwell, and yet I am conscious that a duty devolves upon me to omit no detail.

It was upon the 3rd of May that we reached the little village of Meiringen, where we put up at the Englischer Hof,* then kept by Peter Steiler the elder. Our landlord was an intelligent man, and spoke excellent English, having served for three years as waiter at the Grosvenor Hotel* in London. At his advice, upon the afternoon of the 4th we set off together with the intention of crossing the hills and spending the night at the hamlet of Rosenlaui.* We had strict injunctions, however, on no account to pass the falls of Reichenbach,* which are about half-way up the hill, without making a small detour to see them.

It is, indeed, a fearful place. The torrent, swollen by the melting snow, plunges into a tremendous abyss, from which the spray rolls up like the smoke from a burning house. The shaft into which the river hurls itself is an immense chasm, lined by glistening, coal-black rock, and narrowing into a creaming, boiling pit of incalculable depth, which brims over and shoots the stream onward over its jagged lip. The long sweep of green water roaring for ever down, and the thick flickering curtain of spray hissing for ever upwards, turn a man giddy with their constant whirl and clamour. We stood near the edge peering down at the gleam of the breaking water far below us against the black rocks, and listening to the half-human shout which came booming up with the spray out of the abyss.

The path has been cut half-way round the fall to afford a complete view, but it ends abruptly, and the traveller has to return as he came. We had turned to do so, when we saw a Swiss lad come running along it with a letter in his hand. It bore the mark of the hotel which we had just left, and was addressed to me by the landlord. It appeared that within a very few minutes of our leaving, an English lady had arrived who was in the last stage of consumption.* She had wintered at Davos Platz,* and was journeying now to join her friends at Lucerne,* when a sudden haemorrhage* had overtaken her. It was thought that she could hardly live a few hours, but it would be a great consolation to her to see an English doctor, and, if I would only return, etc., etc. The good Steiler assured me in a postscript that he would himself look upon my compliance as a very great favour, since the lady absolutely refused to see a Swiss physician, and he could not but feel that he was incurring a great responsibility.

The appeal was one which could not be ignored. It was impossible to refuse the request of a fellow-countrywoman dying in a strange land. Yet I had my scruples about leaving Holmes. It was finally agreed, however, that he should retain the young Swiss messenger with him as guide and companion while I returned to Meiringen. My friend would stay some little time at the fall, he said, and would then walk slowly over the hill to Rosenlaui, where I was to rejoin him in the evening. As I turned away I saw Holmes with his back against a rock and his arms folded, gazing down at the rush of waters. It was the last that I was ever destined to see of him in this world.

When I was near the bottom* of the descent I looked back. It was impossible, from that position, to see the fall, but I could see the curving path which winds over the shoulder of the hill and leads to it. Along this a man was, I remember, walking very rapidly. I could see his black figure clearly outlined against the green behind him. I noted him, and the energy with which he walked, but he passed from my mind again as I hurried on upon my errand.

It may have been a little over an hour before I reachd Meiringen. Old Steiler was standing at the porch of his hotel.

'Well,' said I, as I came hurrying up, 'I trust that she is no worse?'

A look of surprise passed over his face, and at the first quiver of his eyebrows my heart turned to lead in my breast.

'You did not write this?' I said, pulling the letter from my pocket. 'There is no sick Englishwoman in the hotel?'

'Certainly not,' he cried. 'But it has the hotel mark upon it! Ha! it must have been written by that tall Englishman who came in after you had gone. He said—'

But I waited for none of the landlord's explanations. In a tingle of fear I was already running down the village street, and making for the path which I had so lately descended. It had taken me an hour to come down. For all my efforts, two more had passed before I found myself at the fall of Reichenbach once more. There was Holmes's Alpine-stock* still leaning against the rock by which I had left him. But there was no sign of him, and it was in vain that I shouted. My only answer was my own voice reverberating in a rolling echo from the cliffs around me.

It was the sight of that Alpine-stock which turned me cold and sick. He had not gone to Rosenlaui, then. He had remained on that three-foot path, with sheer wall on one side and sheer drop upon the other, until his enemy had overtaken him. The young Swiss had gone too. He had probably been in the pay of Moriarty, and had left the two men together. And then what had happened? Who was to tell us what had happened then?

I stood for a minute or two to collect myself, for I was dazed with the horror of the thing. Then I began to think of Holmes's own methods and to try to practise them in reading this tragedy. It was, alas! only too easy to do. During our conversation we had not gone to the end of the path, and the Alpine-stock marked the place where we had stood. The blackish soil is kept for ever soft by the incessant drift of spray, and a bird would leave its tread upon it. Two lines of footmarks were clearly marked along the farther end of the path, both leading away from me. There were none returning. A few yards from the end the soil was all ploughed up into a patch of mud, and the brambles and ferns which fringed the chasm were torn and bedraggled. I lay upon my face and peered over, with the spray spouting up all around me. It had darkened since I left, and now I could only see here and there the glistening of moisture upon the black walls, and far away down at the end of the shaft the gleam of the broken water. I shouted; but only that same half-human cry of the fall was borne back to my ears.

But it was destined that I should after all have a last word of greeting from my friend and comrade. I have said that his Alpine-stock had been left leaning against a rock which jutted on to the path. From the top of this boulder the gleam of something bright caught my eye, and, raising my hand, I found that it came from the silver cigarette-case which he used to carry. As I took it up a small square of paper, upon which it had lain, fluttered down on to the ground. Unfolding it I found that it consisted of three pages torn from his note-book and addressed to me. It was characteristic of the man that the direction was as precise, and the writing as firm and clear, as though it had been written in his study.

'MY DEAR WATSON,' he said, 'I write these few lines through the courtesy of Mr Moriarty, who awaits my convenience for the final discussion of those questions which lie between us. He has been giving me a sketch of the methods by which he avoided the English police and kept himself informed of our movements. They certainly confirm the very high opinion which I had formed of his abilities. I am pleased to think that I shall be able to free society from any further effects of his presence, though I fear that it is at a cost which will give pain to my friends, and especially, my dear Watson, to you. I have already explained to you, however, that my career had in any case reached its crisis, and that no possible conclusion to it could be more congenial to me than this.* Indeed, if I may make a full confession to you, I was quite convinced that the letter from Meiringen was a hoax, and I allowed you to depart on that errand under the persuasion that some development of this sort would follow. Tell Inspector Paterson that the papers which he needs to convict the gang are in pigeon-hole M, done up in a blue envelope and inscribed "Moriarty". I made every disposition of my property before leaving England, and handed it to my brother Mycroft. Pray give my greetings to Mrs Watson, and believe me to be, my dear fellow,

'Very sincerely yours,
'SHERLOCK HOLMES.'

A few words may suffice to tell the little that remains. An examination by experts leaves little doubt that a personal contest between the two men ended, as it could hardly fail to end in such a situation, in their reeling over, locked in each other's arms. Any attempt at recovering the bodies was absolutely hopeless, and there, deep down in that dreadful cauldron of swirling water and seething foam, will lie for all time the most dangerous criminal and the foremost champion of the law of their generation. The Swiss youth was never found again, and there can be no doubt that he was one of the numerous agents whom Moriarty kept in his employ. As to the gang, it will be within the memory of the public how completely the evidence which Holmes had accumulated exposed their organization, and how heavily the hand of the dead man weighed upon them. Of their terrible chief few details came out during the proceedings, and if I have now been compelled to make a clear statement of his career, it is due to those injudicious champions who have endeavoured to clear his memory by attacks upon him whom I shall ever regard as the best and the wisest man whom I have ever known.*

APPENDIX I

The Adventure of the Two Collaborators

At the invitation of Sir James Barrie, Conan Doyle collaborated on the libretto of a comic opera Jane Annie; or, The Good Conduct Prize, *which opened at the Savoy Theatre on 13 May 1893. In* Memories and Adventures, *Conan Doyle wrote*:

Sir James Barrie paid his respects to Sherlock Holmes in a rollicking parody. It was really a gay gesture of resignation over the failure which we had encountered with a comic opera for which he undertook to write the libretto. I collaborated with him on this, but in spite of our joint efforts, the piece fell flat. Whereupon Barrie sent me a parody of Holmes, written on the fly leaves of one of his books. It ran thus:

THE ADVENTURE OF THE TWO COLLABORATORS

In bringing to a close the adventures of my friend Sherlock Holmes I am perforce reminded that he never, save on the occasion which, as you will now hear, brought his singular career to an end, consented to act in any mystery which was concerned with persons who made a livelihood by their pen. 'I am not particular about the people I mix among for business purposes,' he would say, 'but at literary characters I draw the line.'

We were in our rooms in Baker Street one evening. I was (I remember) by the centre table writing out 'The Adventure of the Man without a Cork Leg' (which had so puzzled the Royal Society and all the other scientific bodies of Europe), and Holmes was amusing himself with a little revolver practice. It was his custom of a summer evening to fire round my head, just shaving my face, until he had made a photograph of me on the opposite wall, and it is a slight proof of his skill that many of these portraits in pistol shots are considered admirable likenesses.

I happened to look out of the window, and perceiving two gentlemen advancing rapidly along Baker Street asked him who they were. He immediately lit his pipe, and, twisting himself on a chair into the figure 8, replied: 'They are two collaborators in comic opera, and their play has not been a triumph.'

I sprang from my chair to the ceiling in amazement, and he then explained: 'My dear Watson, they are obviously men who follow some low calling. That much even you should be able to read in their faces. Those little pieces of blue paper which they fling angrily from them are Durrant's Press Notices. Of these they have obviously hundreds about their person (see how their pockets bulge). They would not dance on them if they were pleasant reading.'

I again sprang to the ceiling (which is much dented), and shouted: 'Amazing! but they may be mere authors.'

'No,' said Holmes, 'for mere authors only get one press notice a week. Only criminals, dramatists and actors get them by the hundred.'

'Then they may be actors.'

'No, actors would come in a carriage.'

'Can you tell me anything else about them?'

'A great deal. From the mud on the boots of the tall one I perceive that he comes from South Norwood. The other is as obviously a Scotch author.'

'How can you tell that?'

'He is carrying in his pocket a book called (I clearly see) "Auld Licht Something." Would any one but the author be likely to carry about a book with such a title?'

I had to confess that this was improbable.

It was now evident that the two men (if such they can be called) were seeking our lodgings. I have said (often) that my friend Holmes seldom gave way to emotion of any kind, but he now turned livid with passion. Presently this gave place to a strange look of triumph.

'Watson,' he said, 'that big fellow has for years taken the credit for my most remarkable doings, but at last I have him—at last!'

Up I went to the ceiling, and when I returned the strangers were in the room.

'I perceive, gentlemen,' said Mr Sherlock Holmes, 'that you are at present afflicted by an extraordinary novelty.'

The handsomer of our visitors asked in amazement how he knew this, but the big one only scowled.

'You forget that you wear a ring on your fourth finger,' replied Mr Holmes calmly.

I was about to jump to the ceiling when the big brute interposed.

'That Tommy-rot is all very well for the public, Holmes,' said he, 'but you can drop it before me. And, Watson, if you go up to the ceiling again I shall make you stay there.'

Here I observed a curious phenomenon. My friend Sherlock Holmes *shrank*. He became small before my eyes. I looked longingly at the ceiling, but dared not.

'Let us cut the first four pages,' said the big man, 'and proceed to business. I want to know why—'

'Allow me,' said Mr Holmes, with some of his old courage. 'You want to know why the public does not go to your opera.'

'Exactly,' said the other ironically, 'as you perceive by my shirt stud.' He added more gravely, 'And as you can only find out in one way I must insist on your witnessing an entire performance of the piece.'

It was an anxious moment for me. I shuddered, for I knew that if Holmes went I should have to go with him. But my friend had a heart of gold. 'Never,' he cried fiercely, 'I will do anything for you save that.'

'Your continued existence depends on it,' said the big man menacingly.

'I would rather melt into air,' replied Holmes, proudly taking another chair. 'But I can tell you why the public don't go to your piece without sitting the thing out myself.'

'Why?'

'Because,' replied Holmes calmly, 'they prefer to stay away.'

A dead silence followed that extraordinary remark. For a moment the two intruders gazed with awe upon the man who had unravelled their mystery so wonderfully. Then drawing their knives—

Holmes grew less and less, until nothing was left save a ring of smoke which slowly circled to the ceiling.

The last words of great men are often noteworthy. These were the last words of Sherlock Holmes: 'Fool, fool! I have kept you in luxury for years. By my help you have ridden extensively in cabs, where no author was ever seen before. *Henceforth you will ride in buses!*'

The brute sunk into a chair aghast.

The other author did not turn a hair.

> To A. Conan Doyle.
> from his friend
> J. M. Barrie.

APPENDIX II

HOW I WRITE MY BOOKS
by Sir Arthur Conan Doyle

The following short article by Conan Doyle appeared in What I Think—
a symposium on books and other things by famous writers of
to-day *(ed. H. Greenhough Smith; Newnes, London, 1927). Here Conan
Doyle makes references to some of the problems of 'Silver Blaze'. Although
Conan Doyle did not enjoy such symposia, he appeared at several. He once
told H. Greenhough Smith, the* Strand's *editor, '. . . I hate these Omnium
Gatherum Symposia. I don't see what there is for the Author in them. You
become cheaper the oftener you appear so why make fugitive and honorary
appearances. I have the same objection to charitable scrap-book numbers which
are a perfect plague.'*

When I am asked what my system of work is I have to ask myself
what form of work is referred to. I have wandered into many fields.
There are few in which I have not nibbled. I have written between
twenty and thirty works of fiction, the histories of two wars, several
books of psychic science, three books of travel, one book on
literature, several plays, two books of criminal studies, two political
pamphlets, three books of verses, one book on children, and an
autobiography. For better, for worse, I do not think many men
have had a wider sweep.

In short stories it has always seemed to me that so long as you
produce your dramatic effect, accuracy of detail matters little. I
have never striven for it and have made some bad mistakes in
consequence. What matter if I can hold my readers? I claim that
I may make my own conditions, and I do so. I have taken liberties
in some of the Sherlock Holmes stories. I have been told, for
example, that in 'The Adventure of Silver Blaze', half the characters would have been in jail and the other half warned off the Turf
for ever. That does not trouble me in the least when the story is
admittedly a fantasy.

It is otherwise where history is brought in. Even in a short story
one should be accurate there. In the Brigadier Gerard stories, for
example, even the uniforms are correct. Twenty books of Napoleonic soldier records are the foundation of those stories.

This accuracy applies far more to a long historical novel. It becomes a mere boy's book of adventure unless it is a correct picture of the age. My system before writing such a book as 'Sir Nigel' or 'The Refugees' was to read everything I could get about the age and to copy out into notebooks all that seemed distinctive. I would then cross-index this material by dividing it under the heads of the various types of character. Thus under Archer I would put all archery lore, and also what oaths an archer might use, where he might have been, what wars, etc., so as to make atmosphere in his talk. Under Monk I would have all about stained glass, illumination of missals, discipline, ritual, and so on. In this way if I had, for example, a conversation between a falconer and an armourer, I could make each draw similes from his own craft. All this seems wasted so far as the ephemeral criticism of the day goes, but it is the salt, none the less, which keeps the book from decay. It is in this that Sir Walter Scott is so supreme. I have been reading him again lately, and his work compares to ours as the front of the British Museum to the front of a stuccoed picture palace.

As to my hours of work, when I am keen on a book I am prepared to work all day, with an hour or two of walk or siesta in the afternoon. As I grow older I lose some power of sustained effort, but I remember that I once did ten thousand words of 'The Refugees' in twenty-four hours. It was the part where the Grand Monarch was between his two mistresses, and contains as sustained an effort as I have ever made. Twice I have written forty-thousand-word pamphlets in a week, but in each case I was sustained by a burning indignation, which is the best of all driving power.

From the time that I no longer had to write for sustenance I have never considered money in my work. When the work is done the money is very welcome, and it is the author who should have it. But I have never accepted a contract because it was well paid, and indeed I have very seldom accepted a contract at all, preferring to wait until I had some idea which stimulated me, and not letting my agent or the editor know until I was well advanced with the work. I am sure that this is the best and also the happiest procedure for an author.

EXPLANATORY NOTES

The Memoirs of Sherlock Holmes, containing eleven stories, was published by George Newnes Limited, in an edition of 10,000 copies, as volume three of The Strand Library on 13 Dec. 1893. All the stories were illustrated by Sidney Paget (90 illustrations in all). The first American edition was published on 2 Feb. 1894 by Harper & Brothers, New York. The American edition included 'The Adventure of the Cardboard Box', omitted from the English and subsequent American editions.

SILVER BLAZE

First published in the *Strand Magazine*, 4 (Dec. 1892), 645–60, with 9 illustrations by Sidney Paget, and in the *Strand Magazine*, New York (Jan. 1893). Also published in the USA in: *Baltimore Weekly Sun* (31 Dec. 1892), *Louisville Courier-Journal* (29 Jan. 1893), and *Harper's Weekly*, New York, 37 (25 Feb. 1893), 181–4, with 2 illustrations by W. H. Hyde).

3 *Paddington*: Great Western Railway terminus in 1892.

4 *ear-flapped travelling cap*: This reference follows a similar one in 'The Boscombe Valley Mystery' (*Adventures*) which inspired Sidney Paget (1860–1908), who illustrated the early Sherlock Holmes stories for the *Strand*, to depict Holmes in the deer-stalker hat which has become the detective's trade mark. Paget's brother, Walter, had originally been intended as the illustrator but W. J. K. Boot, the *Strand*'s art editor, sent the commission to Sidney by mistake. Sidney used Walter as his model for the Sherlock Holmes drawings and Walter did, eventually, illustrate one Holmes story, 'The Adventure of the Dying Detective', which appeared in the Dec. 1913 edition of the *Strand*.

the calculation is a simple one: Sherlockian enthusiasts have argued over whether or not this really is a simple calculation. Had Holmes stated the time it took to travel between two telegraph posts, there would be no difficulty. Without this information, however, the accuracy of Holmes' mathematical calculations must remain the subject of conjecture.

the Telegraph and the Chronicle: right-wing and left-wing London newspapers respectively: the *Daily Telegraph* (founded 1855) and the *Daily Chronicle* (founded 1869) were both good pursuers of news, vigorous social commentators, and more reader-conscious than ideological.

5 *young Fitzroy Simpson*: Sir James Young Simpson (1811–70), Edinburgh pioneer of the medical use of chloroform as Professor of Midwifery from 1840. 'Fitzroy' signifies illegitimate royal descent.

'Silver Blaze,' said he, 'is from the Isonomy stock' : ACD appears to have created the horse's name from a combination of Silvio and St Blaise, Derby winners in 1887 and 1883 respectively. Isonomy won the Ascot Cup in 1879 and was a quadruple winner in 1880. The name 'Somomy' appears erroneously in some editions.

6 *One of these lads sat up each night in the stable*: ACD, by his own admission, knew little of racing when he wrote 'Silver Blaze'. 'In and about Newmarket', an article which appeared in the *Strand* (Aug. 1891, the same number as 'The Red-Headed League') may, however, have suggested this particular precaution to him. The head stable-lad is reported as having imparted the following piece of information:

> It was away back to 1875 when Prince Batthyany's Galopin won the Derby. Our friend here had charge of the horse. 'Why, do you know, sir,' he said, 'I slept in the same stall as that horse did for three weeks, so as to make sure that not a living soul got near him; and then when the beauty was sent to Epsom to run in the great race, and win, sir, as I knew he would, although there were a couple of detectives watching, yet I stood outside the stable door all night . . .' (p. 169).

Capleton: Originally Mapleton in the *Strand* (Dec. 1892). This had been changed to Capleton by the time the story appeared in the first book edition of the *Memoirs* in Dec. 1893, probably because Mapleton was a real country seat (though in Kent, not Devon) and its associations in the story are of questionable legality.

a few roaming gipsies: ACD frequently mentions *gipsies* in his stories. His fascination with the race may have stemmed from his love of the writings of George Borrow (1803–81). *Lavengro*

275

and *Romany Rye*, Borrow's own celebrations of the Romany life-style, inspired ACD's own short story 'Borrowed Scenes' (1913). A lengthy appreciation of Borrow's work is included in ACD's *Through The Magic Door* (1907).

7 *furlongs*: two hundred and twenty yards.

touts: horse-race spies frequenting trial runs, racing stables, etc.

10 *a Penang lawyer*: a walking-stick made from palm wood imported from Penang, an island off the west coast of Malaya. The stick was usually made with a bulbous head.

12 *landau*: a four-wheeled carriage, with two facing long seats.

14 *an A.D.P. briar-root pipe*: The initials A.D.P. have previously been accepted to mean 'Alfred Dunhill Pipe'. It is now certain, however, that this is incorrect, as the Dunhill company was not founded until 1907 and did not begin manufacturing pipes util 1910, and the Dunhill pipe with its white spot until 1912. There was a profusion of pipe manufacturers at the time, and the initials would probably have represented a manufacturer's name.

15 *Cavendish*: tobacco that has been flavoured with sugar, maple, or rum, heated and then pressed into cakes to give it a darker colour. The processáproduces a heavy-sweet and mild taste when added to other tobaccos.

cataract knife: a small and delicate knife which is used in the removal of the lens of the eye in cataract surgery. ACD was to investigate a real-life use of a similar knife during his efforts to prove the innocence of George Edalji, who was accused of cattle maiming in the Staffordshire village of Great Wyrley in 1903.

Twenty-two guineas: i.e. twenty-three pounds and two shillings, a guinea being one pound and one shilling sterling. The guinea was favoured as a pricing unit by lawyers, doctors, and smart shops.

20 *'You must see . . .' / '. . . rely upon me'*: these two lines were not in the *Strand*, nor in subsequent American editions.

21 *the apple of his eye*: 'He found him in a desert-land, and in the waste howling wilderness; he led him about, he kept him as the apple of his eye' (Deuteronomy 32: 10).

23 *To the curious incident of the dog in the night-time*: This has, undoubtedly, become one of the best remembered phrases in the whole of the canon, and is perhaps the finest example of the type of epigram which Monsignor Ronald Knox (1888–1957) termed the '*Sherlockismus*' (*Essays in Satire*, 1928). ACD's use of the dog's recognition of his master in this story may be contrasted with the episode in 'Shoscombe Old Place' (*Case-Book*), when Holmes released a spaniel with the intention of revealing a true identity. C. Russell Smith has noted (*Baker Street Journal*, vol. 2, no. 4, 1952) the similarity between the dog in the night-time and an incident in bk. XVI of Homer's *Odyssey*:

> As Telemachus came nearer, the barking dogs did not bark at him; they whimpered round him instead. Odysseus heard the sound of whimpering and the tread of feet that came as well. So his words sped quickly to Eumaeus: 'Eumaeus, surely some friend of yours is about to enter, someone you know at least; the dogs are not barking, they are fawning, and I hear the tread of feet as well'.

drag: carriage drawn by four horses, with seating on top and inside.

24 *off fore-leg*: right foreleg.

Wessex Plate . . . : the winner was to get £50 for each hand (i.e. four inches of the horse's height) plus £1,000. The new course is something over one and a half miles long.

Duke of Balmoral: an inoffensive but obvious way of saying the Prince of Wales (the future Edward VII). After the Royal residence at Balmoral.

Fifteen to five against Desborough! Five to four on the field!: Racing odds are generally reduced to their lowest fraction. However, 6 to 4, 100 o 8, and 100 to 6 are accepted exceptions chiefly, it seems, through tradition. It could reasonably be expected, therefore, that 15 to 5 would be reduced to 3 to 1. The Bookmakers William Hill advise that it is not, however, unusual to hear an on-course bookmaker calling odds of fifteen to five. They add that 'five to four on the field' is also an unusual quotation of odds but that, in days gone by, it would have been possible to back 'the field' against the one horse quoted—in which case the backer would win if any horse other than the favourite came out on top.

25 *spirits of wine*: distilled alcoholic wine solution.

26 *Pullman car*: a railway lounge car or sleeping car of a kind invented by George Pullman (1831–97).

28 *most delicate operations known in surgery*: ACD had attempted eye specialization in early 1891, but wrote Sherlock Holmes stories instead for want of patients.

subcutaneously: beneath the skin.

Editor's Note:

In his autobiography, *Memories and Adventures* (1924), ACD wrote:
Sometimes I have got upon dangerous ground where I have taken risks through my own want of knowledge of the correct atmosphere. I have, for example, never been a racing man, and yet I ventured to write 'Silver Blaze,' in which the mystery depends upon the laws of training and racing. The story is all right, and Holmes may have been at the top of his form, but my ignorance cries aloud to heaven. I read an excellent and very damaging criticism of the story in some sporting paper, written clearly by a man who *did* know, in which he explained the exact penalties which would have come upon every one concerned if they had acted as I described. Half would have been in jail and the other half warned off the turf for ever.

Unfortunately, the article to which ACD referred has never been traced.

THE CARDBOARD BOX

First published in the *Strand Magazine*, 5 (Jan. 1893), 61–73, with 8 illustrations by Sidney Paget, and in (1) *Harper's Weekly*, New York, 37 (14 Jan. 1893), 29–31, and (2) the *Strand Magazine*, New York (Feb. 1893). The story was not included in the first English edition of the series (Newnes, 1893), though it appeared in the first American edition (Harper, 1894). It eventually appeared in its original form in *His Last Bow* (Murray, 1917). Text: *Strand Magazine*.

30 *Parliament had risen*: Parliament had adjourned for the summer recess.

the New Forest: an extensive forested area in Hampshire, which lies south-west of Southampton in the south of England. Originally an ancient hunting forest, it was extended by

William the Conqueror in the eleventh century. It now consists chiefly of oak and beech woodland, coniferous af-forestation, and open heathland. The New Forest was an area in which ACD felt very much at home. *The White Company* (1891) has a New Forest setting and was partly written there; ACD and his family spent many holidays in the area both before and after World War I and, in 1925, he purchased a second home at Bignell Wood. Following ACD's death in 1930, he was buried in the grounds of his home at Crowborough, East Sussex. In 1955, the remains of Sir Arthur and the second Lady Conan Doyle were reinterred in the churchyard at Minstead in the New Forest.

the shingle of Southsea: the beach at Southsea, an eastern suburb of Portsmouth. ACD set up his doctor's practice in Southsea in 1882 and stayed in the town until 1890, when he travelled to Vienna to study the eye. Southsea, therefore, plays an important role in the history of the Sherlock Holmes series of stories for it was during his years there that ACD wrote *A Study in Scarlet* and *The Sign of the Four*.

31   *I read you the passage in one of Poe's sketches*: a reworking of the passage in *A Study in Scarlet*, ch. 2, where Watson had remarked: 'You remind me of Edgar Allan Poe's Dupin. I had no idea that such individuals did exist outside of stories', and Holmes had answered: '. . . Dupin was a very inferior fellow. That trick of his of breaking in on his friends' thoughts with an apropos remark after a quarter of an hour's silence is really very showy and superficial.' The passage which is referred to appears in Poe's *The Murders in the Rue Morgue*. Edgar Allan Poe (1809–49) is considered the undisputed father of the detective story. ACD acknowledged (in *Memories and Adventures*): 'Gaboriau had rather attracted me by the neat dovetailing of his plots, and Poe's masterful detective, M. Dupin, had from boyhood been one of my heroes.' In his much-neglected series of essays on his own literary favourites, *Through the Magic Door*, ACD waxed lyrical on Poe:

Poe is, to my mind, the supreme original short-story writer of all time. His brain was like a seed-pod full of seeds which flew carelessly around, and from which have sprung nearly all our modern types of story. Just think of what he did in his offhand, prodigal fashion, seldom troubling to repeat a success, but pushing on to some

new achievement. To him must be ascribed the monstrous progeny of writers on the detection of crime—
'*quorum pars parva fui!*' [of which things I was a small part. From Virgil, *Aeneid*, II. line 6, where Aeneas tells Dido *quorum pars magna* (great) *fui*, speaking of the Trojan-Hellenic wars]. Each may find some little development of his own, but his main art must trace back to those admirable stories of Monsieur Dupin, so wonderful in their masterful force, their reticence, their quick dramatic point. After all, mental acuteness is the one quality which can be ascribed to the ideal detective, and when that has once been admirably done, succeeding writers must necessarily be content for all time to follow in the same main track. But not only is Poe the originator of the detective story; all treasure-hunting, cryptogram-solving yarns trace back to his 'Gold Bug', just as all pseudo-scientific Verne-and-Wells stories have their prototypes in the *Voyage to the Moon* and the 'Case of Monsieur Valdemar'. If every man who receives a cheque for a story which owes its springs to Poe were to pay a tithe to a monument for the master, he would have a pyramid as big as that of Cheops.

31 *tour-de-force*: a feat of strength, power, or skill.

in rapport: later versions of the story have this phrase in the french style, *en rapport*: in accord, or in agreement.

32 *General Gordon*: General Charles George Gordon (1833–85) served in the Crimean War and then in China, where he earned the nickname Chinese Gordon after suppressing the Taiping Rebellion of 1864. In 1874 he was employed by the Khedive of Egypt to open up the country and, from 1877 until 1880, was British Governor of the Sudan. In 1884 he was sent back to the Sudan to evacuate Europeans and Egyptians, following al-Mahdi's revolt. Gordon was besieged for ten months in Khartoum, where he was murdered after the city was taken two days before a relief force arrived.

Henry Ward Beecher: (1813–87), spell-binding Congregationalist preacher and minister at the Plymouth Church of Brooklyn, NY, from 1847. A zealot in the anti-slavery cause, his support of the 'Free Soil' forces, pledged to keep Kansas territory free from slavery in the mid-1850s, led to the smuggling there of

rifles known as 'Beecher's Bibles'. Beecher, a vigorous supporter of President Abraham Lincoln and the Union cause in the
American Civil War, visited Europe to rally support in
summer 1863. Although the likelihood of Britain going to war
against the Union had ended by 1862, and that of recognizing
the rebellious Confederacy by 1863, aristocratic British
hostility to the United States (utilizing mob support) was too
powerful for Beecher to speak there until after the Union
victories at Gettysburg and Vicksburg on 4 July 1863. He was
faced by concerted efforts to prevent his speaking at Man-
chester (9 Oct.), Glasgow (13th), and above all Liverpool
(16th), where Beecher struggled successfully for three hours to
speak against pro-slavery and pro-Confederacy mob yells,
threats, and barracking. He had a comparatively easy recep-
tion in Edinburgh, where ACD was then nearing the age of
four-and-a-half (but ACD is more likely to have heard
Beecher when he spoke at Portsmouth on 24 Sept. 1886, less
than six months before his death).

A full account of the ordeal, the speeches, interruptions,
and Beecher's ultimate exhaustion and loss of voice in 1863
appeared in ch. xx of William C. Beecher and Revd Samuel
Scoville, assisted by Mrs Henry Ward Beecher, *A Biography of
the Rev. Henry Ward Beecher*, whose 'new and cheaper edition'
was published in London by Sampson Low in 1891 with a fine
engraving of Beecher suitable for framing. This seems the
obvious stimulus as well as source for Watson's meditation in
the story. The reference to Beecher may, however, have
further connotations. First, in his later years Beecher was
accused by Theodore Tilton of having improper relations
with Tilton's wife. As a consequence, Beecher's reputation as
a man of honour and as a clergyman suffered. Secondly,
Harriet Beecher Stowe (1811–96), the author of *Uncle Tom's
Cabin*, was Beecher's sister and it is frequently mentioned by
her biographers that she received a mailed package containing the severed ear of a black man, together with an offensive
note which ascribed to her 'love of niggers' and her impas-
sioned words all the deaths of the Civil War.

the Civil War: the American Civil War (1861–5) was a conflict
between the Federal government and the eleven rebel states
in the south which arose from the conflict of interest between
the predominantly agricultural slave-owning south and the

industrialized north. The election of Abraham Lincoln, a president opposed to slavery, precipitated the secession of the southern states one by one to unify as the Confederate States of America under Jefferson Davis. War broke out when the Confederates opened fire on Fort Sumter, South Carolina, which the Federal government had refused to evacuate. The Federal victory was completed by 26 May 1865 after a long and bloody war.

34 *Daily Chronicle*: a London newspaper, Liberal, moralistic, and reformist, to which ACD had contributed a poem 'HMS *Foudroyant*' (attacking the sale of Nelson's flagship to the Germans for £1,000) published on 12 Sept. 1892.

honeydew tobacco: tobacco sweetened with molasses.

35 *antimacassar*: a covering put over the backs of chairs as protection from grease or as an ornament. Macassar was an oil produced as a hair dressing.

36 *probably a J*: a J pen was a broad-pointed, brass-nibbed desk pen.

thought: later editions changed this to 'meditation'.

37 *carbolic or rectified spirits*: ACD is drawing on his own medical knowledge to provide Holmes with the background information necessary for this particular deduction. Carbolic is disinfectant; rectified spirits are purified alcohol (corpses in the early nineteenth-century Edinburgh medical school were preserved in whisky).

39 *Conqueror ... May Day*: these were two real ships registered under the ownership of the Liverpool, Dublin, and London Steam Packet Co.

he broke the pledge: see n. to p. 48, '*blue ribbon*'.

41 *Stradivarius*: Antonio Stradivari (*c.*1644–1737) was an Italian violin-maker, a pupil of Nicolò Amati (1596–1684), the last great craftsman of the Amati family. From 1666, Stradivari and two of his sons made outstanding violins, violas, and cellos at their workshop in Cremona. Stradivari signed his instruments with the Latin form of his name, Stradivarius.

Jewish broker's: i.e. a pawnbroker whose defaulted client evidently had as little knowledge of the value of what he pawned as did the broker. The word also connotes an authorized seller of distrained goods in settlement of unpaid debt, and this could account for his possession of it, especially if the original owner died before sale. Holmes was lucky: a Jewish broker was likely to know musical artefacts better than would most of his colleagues, except Italians.

Paganini: Nicolò Paganini (1782–1840) was an Italian virtuoso violinist who toured Europe (1828–34) astonishing audiences with his techniques. He composed six violin concertos and various show-pieces for violin, including a set of variations on the G string, and twenty-four caprices, one of which became the basis for compositions by Rachmaninov and Brahms. His last English impresario was called Watson.

claret: red wine from the Bordeaux region of France. Although other countries produce claret-style wines, it is only the *appellation* wines of this region which are correctly known as claret.

42   *That is it*: 'that is the name' in many subsequent versions of the story.

we have been compelled to reason backwards from effects to causes: interestingly, ACD wrote in *Memories and Adventures* (p. 126):

People have often asked me whether I knew the end of a Holmes story before I started it. Of course I do. One could not possibly steer a course if one did not know one's destination. The first thing is to get your idea. Having got that key idea one's next task is to conceal it and lay emphasis upon everything which can make for a different explanation. Holmes, however, can see all the fallacies of the alternatives, and arrives more or less dramatically at the true solution by steps which he can describe and justify.

43   *Anthropological Journal*: properly the *Journal of the Anthropological Institute of Great Britain and Ireland*, a society which was formed in 1871 and to which the prefix Royal was added in 1906. Vol. xx (1891) included the 'Notes' by Dr F. J. Mouet on the paper translated by him on 22 Apr. 1890 from Jacques Bertillon (1851–1922), Président de la Société de statistique de Paris, on 'the Method now practised in France of identifying Criminals'. Bertillon, brother of the more famous Alphonse (1853–1914) — who later told him that Holmes occasionally confused certainty with presumption but that ACD had analytical

genius—was speaking also on behalf of his brother, stressing the unalterable form of the ear in disguise (p. 187). The presiding Vice-President, Sir Francis Galton (1822–1911), was at that time deeply engaged in research on hereditary physical features and traits, and reported frequently on his findings to the Society. Holmes is therefore claiming to be Galton and the Bertillons: a nice reversal of the 'real-life original' ascriptions.

44 *pinna*: the broad upper part of the external ear.

46 *Albert Dock*: part of the system of the Royal Victoria and Albert Docks, which was the largest docking system of London's port, accommodating all but the largest commercial steamers.

darbies: handcuffs.

47 *Shadwell*: a very poor region, near to the docks in London's Stepney district.

plug: hard tobacco, sold for pipe-smoking by the ounce, but also used for chewing. Browner means one chew.

blue ribbon: abstaining from drink. The term derives from the piece of blue ribbon worn in a buttonhole that was the mark of the Blue Ribbon Army, a society pledged to total abstinence from the use of intoxicating drink.

49 *he knew more of the poop than the forecastle*: he was better acquainted with officers' quarters than crewmen's, i.e. Fairbairn may have been a disgraced or bankrupt officer.

50 *hogshead*: a large cask, of liquid or dry measure, usually about fifty imperial gallons in capacity.

started one of our plates: dislocated a sheet of metal in the ship's hull.

51 *New Brighton*: a town on the Wirral peninsula which lies on the opposite bank of the River Mersey to Liverpool. It was a highly popular seaside resort around the turn of the century.

THE YELLOW FACE

First published in the *Strand Magazine*, 5 (Feb. 1893), 162–72, with 7 illustrations by Sidney Paget, and in (1) *Harper's Weekly*, New York, 37 (11 Feb. 1893), 125–7, with 2 illustrations by W. H. Hyde, and (2) *Strand Magazine*, New York (Mar. 1893). John Dickson Carr notes (in *The Life of Sir Arthur Conan Doyle*, 1949) that ACD's diary for 1892 shows that the story was originally entitled 'The Livid Face'.

53 *in*: not in the *Strand*. American edns. have it, but the paragraph is in square brackets and this sentence reads inexplicably 'made us the listeners . . . the actors'.

the affair of the second stain: Watson refers to an adventure which was not to appear until the December 1904 edition of the *Strand*. It was possibly a title, or even a plot, which ACD had planned for some time as it was also mentioned in 'The Naval Treaty'. There is a variation here from the original text which appeared in the Feb. 1893 issue of the *Strand*: there ACD refers to 'The Adventure of the Musgrave Ritual', a story which was to appear in May of the same year. But neither 'The Musgrave Ritual' nor 'The Second Stain' as actually printed are failures.

one of the finest boxers of his weight that I have ever seen: ACD's own love of boxing is revealed in many references to the sport in the Sherlock Holmes stories. He also included a detailed account of the sport in pre-Regency times in *Rodney Stone* (1896) and, on a slightly more supernatural note, wrote possibly the only story which features a hard-hitting boxing ghost, 'The Bully of Brocas Court' (1921).

Save for the occasional use of cocaine: It was possibly a conscious decision to make Holmes appear less dependent on the drug that led ACD to moderate his description of Holmes's addiction from that which appeared in *The Sign of the Four*:

> Three times a day for many months I had witnessed this performance . . .
>
> 'Which is it today,' I asked, 'morphine or cocaine?'
>
> . . . 'It is cocaine,' he said, 'a seven-per-cent solution. Would you care to try it?'

the Park: Regent's Park would have been the Park nearest at hand, but ACD sometimes used the term such that Hyde Park might be meant. The ambiguity arose because he was thinking of The Meadows in Edinburgh, lying between two of his family's residences there, 2 Argyle Park Terrace and 15 Lonsdale Terrace.

54 *our page-boy*: neither style nor syntax link him with Billy ('The Mazarin Stone', *Case-Book*), who sounds mildly Cockney,

where this lad sounds more Irish than anything else. ACD's brother Innes as an 8-year-old acted as his page-boy in the Southsea surgery. The page in Baker Street has other, though non-speaking, appearances.

54 *amber*: a petrified mineralized resin, solidified under centuries of pressure in the ground. Amber is quite fragile and is flammable. It was once the favoured material for pipe mouth-pieces, but its fragility led to an alternative material being substituted.

Why . . . sham amber: 'The American text omits the sentence: I don't know if the editor was shocked that such reprehensible practices should go on among the dishonest English' (Dakin, 98–9).

seven-and-sixpence: seven shillings and six pennies, i.e. thirty-seven and a half new pence.

55 *eightpence*: three new pence.

56 *a brown wide-awake*: a soft broad-brimmed felt hat.

Grant Munro: Sir Alexander Grant (1826–84) was Principal of the University of Edinburgh from 1868 throughout ACD's student years. He is mentioned at the end of *The Mystery of Cloomber*. Alexander Monro was the name of three generations of Professors of Anatomy at Edinburgh (1725–1859). The name was pronounced 'Munro'.

incognito: status as an unknown man.

58 *Atlanta*: the largest town in Georgia, USA, thriving in the 1880s after its destruction by fire in the American Civil War, when it had seceded. Racial segregation there was not fully established until the late 1890s and the early twentieth century.

Jack: i.e. the full name is John Grant Munro.

62 *the Crystal Palace*: a building designed by Joseph Paxton (1801–65), later Sir Joseph, to house the Great Exhibition of 1851 (a display of industrial Britain and Europe, planned by the Prince Consort Albert, which contained some 13,000 exhibits and illustrated the technical progress and industrial supremacy of Britain). It was built in London's Hyde Park with the highly advanced use of prefabricated glass and iron. In 1854, Crystal Palace was dismantled and reassembled at Sydenham, where it burnt down in 1936.

69 *blur*: 'blurr' in *Strand* and Newnes, but later discarded in all edns., presumably as a clear, if curiously persistent, spelling mistake.

71 *little Lucy is darker far than ever her father was*: Sherlockians have raised the anthropological argument that a child of mixed racial marriage has pigmentation approximately half-way between that of the parents.

72 *two*: 'ten' in American edn., presumably because inter-racial marriage was so unacceptable in America in the 1890s, although not in the North in 1880. Despite this emendation 'The Yellow Face' was a lonely banner of freedom on both sides of the Atlantic in the bitter decades ahead.

THE STOCKBROKER'S CLERK

First published in the *Strand Magazine*, 5 (Mar. 1893), 281–91, with 7 illustrations by Sidney Paget, and in (1) *Harper's Weekly*, New York, 37 (11 Mar. 1893), 225–7, with 1 illustration by W. H. Hyde, and (2) *Strand Magazine*, New York (Apr. 1893).

73 *bought a connection*: bought the 'goodwill' of the private practice, which meant retaining any patient willing to remain with the new doctor instead of going to the trouble of looking elsewhere.

St Vitus's Dance: the old name for chorea, involuntary jerky movements, particularly of the hips and shoulders, caused by disease of the part of the brain controlling voluntary movement. It is so called because sufferers in the Middle Ages prayed for a cure at the shrine of St Vitus (the patron saint of dancers).

twelve hundred . . . three hundred: pounds, not patients.

British Medical Journal: so titled from 1857, the chief professional journal of British medicine, attention to which was the hallmark of the conscientious if not absolutely overworked doctor. While still an Edinburgh undergraduate, ACD had a letter, 'Gelseminum as a Poison', which reported on his auto-experimentation, published in its issue of 20 September 1879.

the "Sign of Four": an advertisement for one of ACD's previous novels.

75 *four-wheeler*: an enclosed, four-wheeled cab, drawn by a single horse. Officially known as a clarence cab, unofficially as a growler.

75   *Cockneys*: up to the early seventeenth century, *any* city-dwellers (the term being one of contempt, meaning 'fool', supposedly derived from overly coddled child nursed on 'cocks' eggs' [small, malformed eggs]) and still found thus in the USA. In Britain, London-dwellers and therefore (since persons of wealth lived in country residences outside the London fashionable season) lower-middle, or lower, classes. Distinguished by nasal intonation, flatness of diction, rapidity of speech among lower and middle classes, but in early nineteenth century the famous substitution of 'v' and 'w' for one another, preserved by writings of Charles Dickens (1812–70) notably in *The Pickwick Papers* (1836), proclaimed the Cockney such as Sam Weller and his father. This vanished, but its companion, similar substitution of aspirates for initial vowels and *vice versa*, persisted; these were firmly lower-class and their usage in the middle-class proclaimed a lowly origin to the vigilant. But the Cockney in general, a term quasi-scientifically attached to all persons born within the sound of the bells of St Mary-le-Bow church, was understood to use 'f' or 'v' for 'th' (rationally distinguishing between 'f' for hard and 'v' for soft 'th', respectively in 'thing' ('fing') or 'father' ('faver')). Vowels turning 'a' to 'i', 'e' to 'a' (thus making 'Teddy Taylor' 'Taddy Tyler') were universal around London-dwellers; 'ah' for 'ow' (e.g. 'abaht' for 'about') were held to be lower-class. Watson's use of the term 'class' is significant: middle-class Londoners would be called 'Cockneys', the more easily to place barriers between them and their superiors, and were exposed to snobbish comment, identified with trippers and outings, trade, and vulgarity. Hall Pycroft would not call himself 'All Pycroft, but would be suspected of doing so naturally. But he shared in the more genteel slang of Cockneys in general. It seems to have not too many frontiers with thieves' cant, and while it has been associated with 'rhyming slang' ('trouble and strife' for 'wife') and its derivatives ('butchers' deriving from 'butchers' hook' for 'look'), Henry Bradley opined in the entry on 'Slang' in the *Encyclopaedia Britannica* (1911) that this was the invention of the sporting press. Perhaps the journalists were Cockneys.

Volunteer regiments: these regiments were organized in 1863 to repel any threatened invasion of Britain. By 1890 the number of volunteers stood at a quarter of a million. Legislation was

modified in 1900 to permit volunteer regiments to take part in
the South African War. The regiments became obsolete when
the Territorial Army was created in 1907. ACD was responsible
for reviving the Volunteers, shortly before the outbreak of
World War I, and the new force quickly grew to 200,000 men.

76 *outré*: past participle of French verb 'outrer' (to carry to excess
or beyond reason), hence far-fetched, exaggerated, out-of-the-
way, eccentric, and overdone.

but if I have lost my crib: Cockney slang for 'lost my job'. The
term has quite different meanings in underworld cant, where
it relates to aborted burglary.

let in: slang for deceived.

Venezuelan loan: 'The financial situation in Venezuela was for
a long time extremely complicated and discreditable, owing
to defaults in the payment of public debts, complications
arising from the guarantee of interest on railways and other
public works, responsibility for damages to private property
during civil wars and bad administration ... In 1880–1 there
was a consolidation and conversion of the republic's foreign
indebtedness through a new loan of £2,750,000 at 3%'
(Andrew Jackson Lamoureux, 'Venezuela', *Encyclopaedia Britannica* (1911), 11th edn.). The dictatorship of Guzman Blanco
was overthrown in 1889, which would seem to be the probable cause of the financial convulsion. (By the time the story
was being written, in late 1892, civil war was raging between
the new President, General Andueza Palacios, and an insur-
gent force led by General Joaquin Crespo, which would
ultimately triumph. No doubt ACD was conscious of it, but
the story would seem set in 1889.)

came a nasty cropper: (slang) had a very unpleasant fall.

on the same lay: engaged in the same enterprise (originally
meaning 'criminal enterprise' as used in Charles Dickens's
Oliver Twist (1838)).

Lombard Street: the traditional banking centre of the City of
London which derived its name from Italian merchants who
frequented it prior to the reign of Edward II.

EC: (East Central) the designation of the London postal
district in which most of the stockbroking houses were to be
found.

77   *screw*: slang for wages or salary.

sheeny: a derogatory expression for a Jew. In the first book edition of *The Memoirs* this appeared as Sheeny. In 'The Big Knockover', Dashiell Hammett, writing in *The Black Mask*, Feb. 1927, introduced a rapidly-killed crook named 'Sheeny Holmes'.

78   *Stock Exchange List*: the official daily publication of the Stock Exchange listing the prices of quoted securities.

Ayrshires: railway stock for the Glasgow and South-Western Railway serving Ayrshire.

One hundred and five, to one hundred and five and a quarter: the bid and offer prices quoted by stockbrokers. In the *Strand* text, followed by some American edns. of the story, the prices read 'one hundred and six and a quarter to one hundred and five and seven-eighths'. This is an incorrect quotation as the offer price would always be quoted last, and ACD appears to have been so informed, amending his narrative accordingly.

British Broken Hills: an Australian mining company.

80   *I'll lay you a fiver*: I'll wager five pounds against you.

I took my things to an hotel in New Street: New Street is one of the main streets in Birmingham's city centre, and close at hand to the railway station providing a link with London's Euston terminus.

81   *deal*: cheap produce, pinewood.

83   *Day's Music-Hall*: originally the White Swan public house and then the Crystal Palace, the building was situated in Hurst Street close to Birmingham's city centre. The name was changed to Day's Concert Hall by the owner, James Day, and thus it remained until Day sold the building in 1893. Following the sale, the building underwent yet another change of name, this time to the Empire Theatre. The theatre suffered bomb damage in a 1941 air raid, and was finally demolished in 1951.

very badly stuffed: filled with gold by an obviously incompetent dentist.

comet vintage: wine of a wholly exceptional year, akin in rarity to one in which a major comet passes near Earth.

84   *Arthur Harry Pinner*: 'Arthur Henry Pinner' in Newnes, but this is clearly a misprint. Holmes has no evidence that the name is 'Henry' or was ever given in any form but 'Harry'.

87 *slate-coloured face*: 'clay-coloured face' in the *Strand*. Interestingly, in 'The Red-Headed League' Pope's Court was clogged with men whose hair was variously 'straw, lemon, orange, brick, Irish-setter, liver, clay', the last obviously intended to convey a shade of red not generally conceived as an interpretation of 'clay'. Interesting too is the fact that the villain of 'The Red-Headed League' was one John Clay.

90 *Evening Standard*: founded as the *Standard*, an afternoon paper, in 1827, it became a morning paper in 1857 and acquired its afternoon stable-mate in 1859, both booming in circulation under the management of William Mudford (1839–1916), an independent but firm Tory. It declined after Mudford's retirement in 1899, lost its morning paper in 1916, but thrives today.

cracksman: a burglar or housebreaker (underworld slang). ACD's brother-in-law, Ernest William Hornung (1866–1921), introduced his criminal imitation of Holmes, Raffles, in *The Amateur Cracksman* (1899).

City Police: the square mile of the City of London, containing the capital's financial district, and simply known as 'the City', has a police force of its own. The Metropolitan Police ('Scotland Yard') have no jurisdiction in the City.

affection: Jews were widely credited with exceptional family affection (as is very beautifully expressed, for instance, in Anthony Trollope's *The Way We Live Now*, ch. 79). Sometimes this was made a ground of anti-Semitic derision, but ACD clearly admired it. For sibling affection for a murderer in a non-Jewish context see *The Hound of the Baskervilles*, ch. 13.

THE 'GLORIA SCOTT'

First published in the *Strand Magazine*, 5 (Feb. 1893), 395–406, with 7 illustrations by Sidney Paget, and in (1) *Harper's Weekly*, New York, 37 (15 Apr. 1893), 345–7, with 2 illustrations by W. H. Hyde, and (2) *Strand Magazine*, New York (May 1893).

ACD had a strong interest in seafaring tales and had served as ship's doctor on voyages both to the Arctic and the West African Coast. Many of his own particular favourite tales are mentioned in *Through the Magic Door* (1907). His own collection, *Tales of Pirates and Blue Water* (1922), illustrates his interest in the subject and, amongst others, contains four stories detailing the career of the pirate

Captain Sharkey. In addition, ACD wrote an account of 'The Voyage of the *Flowery Land*' (*Uncollected Stories*), which detailed the circumstances of the penultimate mutiny aboard a British ship. This was possibly intended as a contribution to the unsuccessful series of true crime studies, 'Strange Studies from Life', which appeared in the *Strand* in 1901.

Donald A. Redmond, in *Sherlock Holmes a Study in Sources* (1982), suggests that the plot of 'The "*Gloria Scott*"' was based on the hijacking, or piracy, of the brig *Cyprus* in August 1829. The brig had sailed in mid-August 1829 from Hobart Town in Tasmania, bound for the penal settlement at Macquarie Bay, with some forty convicts guarded by soldiers under a lieutenant. In an incautious moment, while the ship was becalmed, the officer, ship's mate, and some sailors took to a boat to fish. Nine or ten convicts were left on board in chains, but they were able to overpower the sentries and release their fellows. Under threat that the ship would be burned, the crew and military guard surrendered, and were set ashore in Research Bay. Nineteen convicts seized the brig, but twelve of the prisoners refused to join in and were also landed with the crew. Two of the marooned group, Morgan and Pobjoy, managed to row a makeshift boat some twenty miles to Partridge Island, where they were rescued by the *Orilea*. The master of the *Orilea* was named Hudson.

92 *Head-keeper Hudson*: 'Head-keeper' in all texts, but the code requires it as two words, as for 'fly paper' here ('fly-paper' in the *Strand*). ACD's hyphen was often difficult to distinguish from a meaningless pen-stroke.

93 *during the two years that I was at college*: Holmes makes reference to his time at college in two succeeding stories: 'The "*Gloria Scott*"' and 'The Musgrave Ritual', and the location of his college has been the subject of discussion amongst Sherlockians for many decades. It is inappropriate to prolong that discussion here, but a lengthy dissertation may be found in Dorothy L. Sayers's *Unpopular Opinions* (1946).

freezing: underworld slang for 'to render immobile' with a sense of retention, an interesting if somewhat unusual instance of Holmes's vocabulary suffering from criminal contagion. The incident of Victor Trevor's bull-terrier freezing on to Holmes's ankle may have been suggested to ACD by a recent contribution of Arthur Morrison (1863–1945) to the *Strand* (Feb. 1892). In 'My Neighbour's Dogs' Morrison wrote,

'Perhaps while I am praising this very bull-dog, he is filling that immense mouth of his from some inoffensive person's leg. But, as the leg isn't mine, and I hate unreasonable grumbling, I won't grumble about that.' Morrison's 'Martin Hewitt, Investigator' series of detective stories began in the *Strand* in Mar. 1894, three issues after 'The Final Problem'.

J.P.: Justice of the Peace, a lower magistrate appointed for a specific district, administering summary justice in lesser cases, granting licences, etc.

Broads: a low-lying area in eastern England, mainly in Norfolk but extending also into Suffolk, consisting of a system of shallow lakes, believed to have originated as medieval peat diggings, linked to the Rivers Bure and Yare.

94 *a very handsome stick . . . By the inscription*: a deduction by Holmes, reminiscent of the opening scene in *The Hound of the Baskervilles* when Holmes and Watson make their own deductions about the identity of the owner of a Penang lawyer which had been left behind by Dr Mortimer.

95 *callosities*: skin-thickenings or callouses.

finger glasses: water vessels to clean fingers at the dinner table.

96 *governor*: (pronounced 'guv'nah') Victorian aristocratic or squirearchic filial usage to denote father, a strongly upper-class affectation.

97 *harness cask*: a covered cask or tub, usually kept on the deck of a ship, and containing the salted provisions for the day's consumption.

two-yearer in an eight-knot tramp: two years' service on a cargo ship which ran no regular route or schedule and which was capable of a speed no faster than eight knots, or nautical miles per hour.

98 *dog-cart*: a light sporting vehicle with two seats placed back to back. The cart derives its name from the box under the rear seat which was originally used for carrying dogs.

Apoplexy: a rush of blood to the brain resulting in instant death, or paralysis frequently followed by rapid decease.

99 *Broads*: 'the Broads' in the *Strand*.

100 *Fordingbridge*: west Hampshire town on the River Avon, 14 miles south of Salisbury.

101 *Japanese cabinet*: black lacquered cabinet, Japanese style.

103 *barque*: a three-masted sailing ship.

104 *sentenced to transportation*: the practice of sending a convicted criminal to a place outside Britain, usually one of the colonies, where he would be subjected to hard labour. Transportation dates back to the reign of Elizabeth I, and became increasingly popular as a means of providing labour in the colonies towards the end of the 17th century. It was directed towards Australia from the late 18th century and was not abolished for Western Australia until 1868.

105 *when the Crimean War was at its height*: the Crimean War began in 1854 and ended with the Treaty of Paris in Feb. 1856.

the new clippers: extremely fast sailing ships developed by American shipbuilders from 1840 onwards.

106 *kiss the Book*: swear evidence (as in court), using the Bible which was held, and frequently kissed.

the dibbs: slang for money. The spelling 'dibs' is more usual, and occurs in some recent English edns.

107 *keel to maintruck*: bottom to top, 'maintruck' being the cap at the top of the mast.

Bay: the Bay of Biscay.

tracts: printed religious discussions or exhortations to improve one's life, spiritually rather than materially.

108 *head on the chart of the Atlantic*: American texts, departing (unusually for them) from that of the *Strand*, read: 'with his brains smeared over the chart of the Atlantic, on which Dakin (p. 109) commented 'it seems a needlessly horrid alteration'. It may be ACD's original version: doctors are not squeamish.

109 *junk*: salted meat carried on long voyages.

110 *the painter*: the rope attached to the bow of a boat.

foreyard aback: the lowest spar extending the sail's lower part, drawn back to keep clear of wind and slow down the ship.

Cape de Verds: Cape Verde islands.

112 *the Terai*: a swampy lowland belt in India, north of the Ganges River at the foot of the Himalayas.

THE MUSGRAVE RITUAL

First published in the *Strand Magazine*, 5 (May 1893), 479–89, with 6 illustrations by Sidney Paget, and in (1) *Harper's Weekly*, New York, 37 (13 May 1893), 453–5, 458, with 2 illustrations by W. H. Hyde and (2) *Strand Magazine*, New York (June 1893).

It seems likely that ACD took the name Musgrave from two small villages in Westmorland, Little Musgrave and Great Musgrave. The villages are situated a few miles north of Kirkby Stephen and close to the Masongill home of his mother, Mary Doyle, who lived on the estate of Bryan Charles Waller. This was not the only occasion on which ACD drew from names in the Masongill area. In his short story, 'The Surgeon of Gaster Fell' (1890), Gaster Fell is obviously the Casterton Fell which lies within short walking distance of Masongill.

113   *rough-and-tumble work in Afghanistan*: Watson is referring to his work as assistant surgeon with the Fifth Northumberland Fusiliers and the Berkshires during the Afghan War of 1878–80 (see *A Study in Scarlet*).

hair-trigger: an old-fashioned firearm having a secondary trigger which is very delicately adjusted. A very slight pressure upon the hair-trigger would release the main trigger and fire the gun. It was considered suitable only for target- and trick-shooting.

Boxer cartridges: a form of ammunition perfected by Colonel Edward Mourrier Boxer (1822–98).

V.R.: Victoria Regina. The implication would seem to be that this was a celebration of the Golden Jubilee of her accession to the throne, i.e. that the shooting took place on or around 21 June 1887.

somewhere in these incoherent memoirs: see *A Study in Scarlet*, ch. 2.

114   *have*: 'had' in *Strand*, American, and even Baring-Gould edns.

Tarleton: small town in West Lancashire between the seaside at Southport and Stonyhurst College, where ACD was at school.

Ricoletti of the club foot: an idea possibly worked out (minus the abominable wife) in 'The Club-Footed Grocer' (*Strand*, Nov. 1898; reprinted in *Round the Fire Stories* (1908)).

recherché: out of the ordinary. A modern usage would be 'choice'.

116 *Montague Street*: at the end of 1890, ACD decided to leave his home in Southsea to journey to Vienna where he would study the eye. In the spring of 1891, he returned to London and rented rooms in Montague Place (the northern boundary of the British Museum; Montague Street is the eastern). It was from here that he searched for suitable rooms in which to set up his oculist's practice—a profession which was unsuccessful and convinced him that his living should be made from full-time writing.

mullioned windows: windows divided by vertical shafts.

keep: stronghold, usually towering, against invasion, peasants, etc.

117 *I am member*: I am the elected Member of Parliament. This would have been before the Reform Act of 1884 which abolished such 'districts' by greatly enlarging the number of voters in the lower classes.

I preserve: preserve game or fish by raising and protecting it for private use. It is somewhat ironic that the preservers then spend certain months of the year destroying what they have raised and protected.

pheasant months: the shooting of pheasant is legal in Great Britain between Oct. and the end of Jan.

118 *Don Juan*: the great aristocratic libertine of European literature, supposedly based on Don Juan de Tenorio of Seville. His probable first appearance was in Tirso de Molina's play *El burlador de Sevilla* (?1630). The plot of this play is retained in subsequent versions, most notably Molière's *Don Juan* (1665), and Mozart's *Don Giovanni* (1787). He receives more satirical and more affectionate treatment in Byron's unfinished epic *Don Juan* (1819–24), and George Bernard Shaw's *Man and Superman* (1903).

we: if not a royal 'we', something comparable. There are no other Musgraves in residence, so far as we know. Reginald Musgrave at points seems a sharply satirical portrait.

café noir: black coffee.

blazonings and charges: a heraldic term to describe a coat of arms and any device borne upon it.

122 *and the outhouses*: the *Strand* and subsequent texts have 'from cellar to garret'.

building: 'house' in the *Strand* and subsequent edns.

mere: a lake or pond.

123 *'Whose was it?' 'His who is gone.'*...: when 'The Musgrave Ritual' was first published in the *Strand* for May 1893, the third couplet: 'What was the month?' 'The sixth from the first' was missing. This had been added by the time the story was published in the *Memoirs* in Dec. 1893, and has been retained in subsequent British edns., although the American Doubleday edn. (reproduced as the *Penguin Complete Sherlock Holmes*) perpetuates the omission.

T. S. Eliot (1888–1965) utilized the wording of 'The Musgrave Ritual' in his play *Murder in the Cathedral* (1935). Upon Thomas Becket's arrival at Canterbury, the second of four Tempters urges him to submit to King Henry's will, to recover the Chancellorship, and to exercise temporal power for the good of the Kingdom. The stichomythic exchange with Becket runs as follows:

THOMAS
Who shall have it?
TEMPTER
He who will come.
THOMAS
What shall be the month?
TEMPTER
The last from the first.
THOMAS
What shall we give for it?
TEMPTER
Pretence of priestly power.
THOMAS
Why should we give it?
TEMPTER
For the power and the glory.

In 1951 a lively correspondence appeared in the *Times Literary Supplement* (19, 26 Jan., 23 Feb.) following suggestions that Eliot had plagiarized ACD. The argument was settled by Mr Nathan Bengis of New York, whose letter appeared on 28 Sept. 1951:

Sir,—Early this year a spate of letters appeared in these columns concerning Mr T. S. Eliot's use of the Musgrave Ritual (from Sir Arthur Conan Doyle's Sherlock Holmes story of that name) in his *Murder in the Cathedral*. There has been much speculation about this, and it has even been suggested that Mr Eliot and Sir Arthur borrowed the Ritual from a common source.

Remembering Sherlock Holmes's warning about the danger of theorizing before one has all the evidence, I wrote to Mr Eliot in May of this year and asked him about the matter point-blank. I quote with permission from his reply: '... My use of the "Musgrave Ritual" was deliberate and wholly conscious.'

This definitive answer should, I think, end the discussion of this much-mooted point.—Nathan L. Bengis, Keeper of the Crown, the Musgrave Ritualists of New York.

Professor W. W. Robson remarks that ACD made no great claim for his poetry, but that the Ritual is fine poetry in itself, and better than Eliot's version.

124 *ten generations of his masters*: 'Holmes's unkind but justifiable comparison ... is ... rather surprising: one would have expected the interval 1660–1879 to comprise seven generations at most. Perhaps he credited them with faster breeding than thinking' (Dakin, p. 116).

131 *These are coins of Charles I*: in a numismatic sense, the reign of Charles I (1625–49) was one of the most interesting of all the English monarchs. Some machine-made coins were produced by Nicholas Briot, a French die-sinker, but they could not be made at the speed of the previous hand-hammering method. In 1637 a branch mint was set up at Aberystwyth to coin silver extracted from the Welsh mines. Following Charles's final breach with Parliament, the parliamentary government continued to issue coins bearing his name and portrait until his trial and execution.

During the Civil War, coins were struck at a number of towns to supply coinage for those areas of the country under Royalist control. Among the more spectacular pieces are the gold triple unites and the silver pounds and half-pounds struck at Shrewsbury and Oxford, and the emergency coins made from oddly shaped pieces of silver plate during the sieges of Newark, Scarborough, Carlisle, and Pontefract. The silver and copper coinage of the time would hardly have deteriorated into the 'three rusty old discs of metal' which Watson describes at the beginning of this story. One can only assume that Conan Doyle chose the wrong adjective for his description.

132 *Cavalier*: a supporter of the Royalist party during the English Civil War. Cavaliers were distinguished by their elaborate dress, with lace ruffles, feathers, and velvet, which contrasted with the sober attire of the parliamentary Roundheads.

the ancient crown of the Kings of England: possibly the ancient Crown of St Edward which 'once encircled the brows' of two royal Stuarts: James I (1566–1625) and Charles I (1600–49), on the occasion of their respective coronations. The 1649 inventory of British regalia described the crown of St Edward (listed as King Alfred's Crown) as being of 'gould wyer worke, sett with slight stones'.

Charles II, whose advent was already foreseen: following the fall of the Protectorate in 1659, General George Monck (1608–70), afterwards 1st Duke of Albemarle, negotiated (1660) the Restoration of the monarchy and Charles (1630–85) became king after promising a general pardon and liberty of conscience in the Declaration of Breda. But such Royalists as the Musgraves would have preferred to prophesy a Restoration by force of Cavalier arms without dependence on a pragmatic Cromwellian.

133 *a considerable sum to pay*: Bengis questioned the idea of the crown being sold (Baring-Gould, i. 140) as being 'a public possession, and therefore no more negotiable than London Bridge'. The auctorial intention here was indeed probably satirical, given his poem 'HMS *Foudrayant*' (*Daily Chronicle*, 12 Sept. 1892), a bitter denunciation of the sale of Nelson's flagship, whose second verse runs:

> There's many a crypt in which lies hid
>   The dust of statesman or of king;
> There's Shakespeare's home to raise a bid,
>   And Milton's house its price would bring.
> What for the sword that Cromwell drew?
>   What for Prince Edward's coat of mail?
> What for our Saxon Alfred's tomb?
>   They're all for sale!

THE REIGATE SQUIRE

First published in the *Strand Magazine*, 5 (June 1893), 601–12, with 7 illustrations by Sidney Paget, and in (1) *Harper's Weekly*, New York, 37 (17 June 1893), 574–6, with 2 illustrations by W. H. Hyde, and (2) *Strand Magazine*, New York (July 1893). On its first publication in the *Strand* this story was entitled 'The Adventure of the Reigate Squire'. Richard Lancelyn Green has noted (in *Baker Street Dozen*) that it seems likely the original working title was 'The Reigate Puzzle', as this was the name used by the illustrator, Sidney Paget, in his account book in Mar. 1893. 'The Reigate Puzzle' has remained in use in the United States, but the story's title had changed to 'The Reigate Squires' by the time the Newnes first edition of the *Memoirs* appeared in Dec. 1893. If we accept the *Oxford English Dictionary*'s definition of the word 'squire' as meaning, especially, one who is the chief land-owner, magistrate, or lawyer in a district, then Cunningham Senior, who was the JP, becomes the squire in question and it follows, therefore, that the *Strand*'s original title is the more correct of the two British variants.

134   *Baron Maupertuis*: this is misspelled as Maupertins in the story's original *Strand* appearance, but corrected in all other texts. The name is doubtless taken from the French mathematician, astronomer, and pioneer geneticist, Pierre Louis Moreau de Maupertuis (1698–1759), sent by Louis XV to the Gulf of Bothnia in July 1736 to measure an arc near the polar circle of 57° amplitude as part of a scheme to measure the earth. He later schemed to establish a genetic system capable of producing all imaginable varieties across the gamut of possibilities, which inspired H. G. Wells's *The Island of Dr Moreau* (1896). He became locked in bitter enmity against Voltaire (1694–1778), a slightly Holmesian figure whose *Zadig* (1747) is sometimes cited as one of the first fictional treatments of detection.

  Lyons: the capital of the Rhône department and the third largest city in France, Lyons lies at the confluence of the Rivers Rhône and Saône. The city is an important financial centre and has been a leading textile centre since the fifteenth century.

135   *Reigate*: a town in Surrey, south-eastern England, at the foot of the North Downs. It is now a dormitory town for London.

  Pope's 'Homer': Alexander Pope (1688–1744) issued the first volume of his translation in heroic couplets of Homer's *Iliad* in 1715. The work was completed in 1720 and was supplemented in 1725–6 by his translation of the *Odyssey* in which he was assisted by William Broome and Elijah Fenton.

136 *J.P.*: Justice of the Peace. This magistrate is authorized to sign documents requiring official signature, such as warrants for search or arrest.

137 *crack two cribs*: underworld slang. To 'crack a crib' is to commit a burglary.

138 *Mr Cunningham ... Mister Alec*: implies 'Mr' is pronounced differently from 'Mister', the latter being a courtesy title like the Viscountcy accorded to an Earl's son. In Scotland, whence the name Cunningham, ACD was used to 'Maister' as a salutation to a patriarch of property (see 'The Man from Archangel', in *The Captain of the 'Pole-Star'*).

140 *method in his madness*: '*Insanire paret certa ratione modoque*' (He may prepare to go mad with certain reason and method), Horace, *Satires*, II.iii.271; '*Polonius*: Though this be madness, yet there is method in't' (Shakespeare, *Hamlet*, II.ii.208); 'My friend, about whose madness I now saw, or fancied that I saw, certain indications of method' (Poe, 'The Gold-Bug').

142 *the date of Malplaquet*: the Battle of Malplaquet took place on 11 Sept. 1709 during the course of the War of the Spanish Succession. The French faced the armies of Britain, the Netherlands, and Austria under the command of Marlborough and Prince Eugene of Savoy and fought the battle some ten miles south of Mons. By a strategic retreat, the French inflicted severe casualties upon the allies, who, although victorious, were checked in their advance upon Paris.

148 *stormy petrel*: a common oceanic seabird of ill omen, thought by sailors to foretell an approaching storm.

151 *the deduction of a man's age from his writing ... by experts*: the use of graphology as a plot device is a recurring theme in the Sherlock Holmes series of stories but the importance of graphology as a means of deduction is given most emphasis in 'The Reigate Squire'. In the 19th century, Abbé Flandrin, Abbé Michon, and Desbarolles pioneered the analytical system of graphology in France by establishing a technique of fixed signs. In the early 1880s, Crepieux-Jamin observed that handwriting is, in a circumscribed, subtle, and infinitely

variable way, gesture, and that these gestures can express changing moods, emotion, and bodily well-being. Crepieux-Jamin's theory was scientifically tested by others, including Alfred Binet (1857–1911), who pioneered the principles used in intelligence tests. Systems of graphology developed from it and the science is now commonly used, particularly by personnel assessors. ACD was to draw on graphology to support his arguments to prove the innocence of George Edalji during his investigations in 1907.

Richard Lancelyn Green notes (see *Baker Street Dozen*) that ACD probably had the idea for the basis of this story suggested to him. During the Christmas holidays of 1892, Alexander Cargill, a handwriting expert from Edinburgh, offered to send ACD his article 'Health in Handwriting', which had been published in the *Edinburgh Medical Journal* (Jan. 1890, pp. 627–31, with eight illustrations), as he hoped this might provide material for a story. Cargill admitted that chirography was an uncertain art, but his essay showed that an expert could distinguish between scripts. Not only could physical characteristics be discovered, but also the age of the writer could be reasonably ascertained. Cargill had no medical degree and was the only such contributor to vol. 35 (July 1889–June 1890): he was otherwise a local sonneteer in Scots and English (*Songs from a Pedlar's Wallet* (1883)), and sententiary (*Work-a-Day Thoughts* (1886)). It may be noted that Joseph Bell had contributed 'Note on a Case of Loss of Memory following Cranial Injury' to the *Edinburgh Medical Journal* (which he edited) the previous month (p. 543).

Cargill's specimens included an octogenarian (who omitted some 'i'-dots), 'normal healthiness in the writing of an insane person' and 'decided evidences of an alcoholic's insanity' (with a few touches reminiscent of Alec Cunningham's hand), and 'Person suffering from cerebral disease (earlier stage)' suggesting both Cunninghams. Holmes's phrase 'his son was a perfect demon' was, in fact, intended to have a medical significance. ACD may have seen Cargill's article before its author wrote to him, being still in active practice in Jan. 1890 and likely to read his old medical school's chief publication. Its ironies in the context of his own father are terrible.

ACD returned thanks for the article to Cargill on 6 Jan. 1893, writing:

I am almost afraid to write to you, for fear you should discover imbecility in the dots of my i's, or incipient brain softening in my capitals. You have given me quite new ideas, and I thank you for them, and can recognize their truth since they tally with every man's experience, tho' I never saw them condensed and formulated before. I would like now to give Holmes a torn slip of a document, and see how far he could reconstruct both it and the writers of it. I think, thanks to you, I could make it effective.

152 *a single paper*: this accounts for the choice of the name 'Acton'. The professionalization of historiography and its dependence on the evidence of documents was particularly associated with Sir John Acton (1834–1902), first Lord Acton, shortly to become Regius Professor of modern History at Cambridge. He was known to have contemplated burglary to discover crucial historical documents.

THE CROOKED MAN

First published in the *Strand Magazine*, 6 (July 1893), 22–32, with 7 illustrations by Sidney Paget, and in (1) *Harper's Weekly*, New York, 37 (8 July 1893), 645–7, with 2 illustrations by W. H. Hyde, and (2) *Strand Magazine*, New York (Aug. 1893).

155 *carrying your handkerchief in your sleeve*: Holmes refers to a habit Watson would have acquired during his time in the army: the uniform tunic had no pocket for carrying a handkerchief.

British workman: the expression derives from the official term 'British plateworker' for persons legally qualified to work in pewter plate. ACD is using the term in the form of the popular joke that similar expertise is attested for sewage or gas maintenance men, where inexpertise was reputedly the norm.

156 *hansom*: a two-wheeled covered carriage drawn by a single horse, named after its patentee J. A. Hansom (1803–82). The hansom held two people in addition to the driver, who was mounted upon an elevated seat behind the body of the carriage.

Elementary: one of the two occasions in the Holmes series when Holmes uses the phrase 'elementary'. This is as near as ACD came to writing the much quoted apocryphal phrase 'Elementary, my dear Watson' which never appeared in the stories.

157   *the Royal Mallows*: a non-existent regiment thought, by some, to be the second battalion, the Royal Irish Fusiliers. The name may also have been inspired by the Royal Munster Fusiliers. American edns. actually call it 'the Royal Munsters', which looks like an original title later changed, perhaps for legal reasons. ACD used the same regimental name in his famous story 'The Green Flag', which was first published in June 1893, the month prior to the first appearance of 'The Crooked Man'. The name links the regiment with County Cork, one of whose major northern towns is Mallow, disfranchised in the Reform Act of 1884: its last by-election, in 1883, had gained celebrity by choosing the fiery nationalist journalist William O'Brien after the tactically aborted candidacy of ACD's schoolmate, the devious lawyer John Francis Moriarty.

Crimea and the Mutiny: the Crimean War (1853–6), between Russia, on the one side, and Britain, France, the Ottoman Empire, and (after 1855) Sardinia–Piedmont on the other, was caused by Russia's expansionist ambitions in the Balkans. War was precipitated by Russia's desire to establish a protectorate over Orthodox Christians in the Ottoman Empire. Russia occupied Moldavia and Walachia in July 1853; Turkey declared war in October of the same year and, following the destruction of a Turkish fleet at Sinope, Britain and France entered the war in Mar. 1854. The major battles of the year-long siege of Sevastopol were Balaclava and Inkerman. Russia eventually evacuated the port in Sept. 1855 and peace was formally concluded at Paris in 1856. Over 250,000 men were lost by each side, many from disease in the appalling Crimean hospitals, and it was during the campaign that the work of Florence Nightingale was to come to the fore in improving conditions.

The Indian Mutiny (1857–9) was a revolt by some 35,000 sepoys (Indian soldiers in the service of the British East India Company), which developed into a bloody Anglo-Indian war. It began with a massacre of Europeans at Meerut in May 1857. British reinforcements, under their commander-in-chief Colin Campbell, regained Delhi and relieved Lucknow in late 1857. By July 1858 the revolt had been more or less contained.

Devoy: a deliberately ironic choice of name. John Devoy (1842–1928), a violently Anglophobe Irish-American conspirator, started as a Fenian organizer in the British army. ACD is enjoying the use of the name for purposes of British regimental romance.

colour-sergeant: a sergeant having control of battalion or regimental colours.

158 *he had been sunk*: the *Strand*'s first publication of the story in July 1893 has this phrase printed as 'he has been sunk'.

159 *Guild of St George*: based on the St Vincent de Paul Society, whose branch affiliated to St Mary's Roman Catholic Cathedral in Edinburgh, included ACD's mother and father before their marriage.

160 *and his wife*: 'and of his wife' in the *Strand*.

166 *Hudson Street*: the allusion is symbolic, being to Henry Hudson who, in 1611, perished in what is now Hudson's Bay, having been set adrift there by his comrades to die.

167 *registration agent*: one who assists in making up lists of eligible voters for review (this being before universal suffrage)

his box: 'that box' in the *Strand*.

florin: in 1849, as a first, short-lived, step towards decimalization of the British currency, a silver florin (one-tenth of a pound) was introduced. The coins of 1849 omitted the usual *Dei Gratia* from the inscription and became known as 'Godless' florins. It seems unlikely that it is a 'Godless' florin which is referred to in the story, however, and it is more likely that the landlady meant a forged coin.

Indian rupee: the standard silver coin of British India.

168 *Baker Street boys*: Holmes's unofficial police force, the Baker Street Irregulars. It is strange that we only ever encounter boys amongst the street Arabs described by Holmes and Watson, as the appalling poverty of the lower classes in late Victorian London would have ensured that large numbers of girls would also have found themselves on the streets.

a small street Arab: a neglected or abandoned boy or girl of the streets.

170 *cantonments*: a permanent military station forming the nucleus of the European quarter of an Indian city. 'Bhurtee' seems to

be meant for Allahabad, of which ACD would have been given a vivid account by his Southsea patient and mentor, Major-General Alfred Wilks Drayson, who had reported on its defences twenty years after the Indian Mutiny.

170 *General Neill*: General James George Smith Neill (1810–57) who relieved the besieged fort at Allahabad on 11 June 1857 and later commanded the right wing in the advance from Cawnpore to Lucknow during the Indian Mutiny. He was shot and killed during the attack on Lucknow.

171 *Nepal*: Nepaul in the *Strand* and other early editions.

Darjeeling: a military station for the British in India situated in the hill region of northern Bengal.

Punjab: a province of British India, now in Pakistan.

173 *the small affair of Uriah and Bathsheba*: as Holmes points out, the reference is to be found in 'the first or second of Samuel'. In fact it is found in 2 Samuel, 11–12, where it is described how David, king of Israel, coveted Uriah's wife, Bathsheba, who became pregnant while her husband was away fighting a war. David called Uriah home and then sent him off to war again, to a more dangerous part of the battle where he was sure to be killed.

THE RESIDENT PATIENT

First published in the *Strand Magazine*, 6 (Aug. 1893), 128–38, with 7 illustrations by Sidney Paget, and in (1) *Harper's Weekly*, New York, 37 (12 Aug. 1893), 761–3, with 2 illustrations by W. H. Hyde, and (2) *Strand Magazine*, New York (Sept. 1893). Text: *Strand Magazine*.

174 *tour-de-force*: a feat of strength or skill.

Scylla and Charybdis: Homer's *Odyssey* (Book XII) records the two sea-monsters living in caves by the sea. Scylla was renowned for snatching seamen from passing ships whilst Charybdis displayed the particular talent of swallowing the waters, and the ships thereon. *A Study in Scarlet* (involving two murders in revenge for a rape) is curiously categorized as 'slight' or 'commonplace': perhaps this is a reply to Holmes's charge in *The Sign of the Four* of Watson's having romanticized it.

I cannot be sure of the exact date: this paragraph, and the one following it as far as the words 'the wind has fallen', were removed from the story when it was published in the *Memoirs* in Dec. 1893. The reason for this was that a large piece of text which originally appeared in the *Strand*'s version of 'The Cardboard Box' was inserted in its place ('The Cardboard Box' having, by this time, been suppressed by ACD). The John Murray *Complete Sherlock Holmes Short Stories* (1928) dropped the additional matter duplicated in 'The Cardboard Box' (from 'Our blinds were half-drawn' (the latter's para. 2, sent. 3) to 'It was very superficial . . . incredulity the other day,' (in its para. 20), substituting: 'It had been a close, rainy day in October. "Unhealthy weather, Watson," said my friend. "But the evening has brought a breeze with it. What do you say to a ramble through London?"' The American Doubleday edn. (and hence the Penguin *Complete Sherlock Holmes*) retained the thought-reading passage in both short stories.

175 *nose-high*: wrapping his muffler or scarf over his mouth to his nostrils, since as an invalid recently returned from India, Watson feared a chilly climate.

Fleet Street: one of the chief approaches to the City of London, Fleet Street was, until relatively recently, the centre of Britain's newspaper industry.

176 *Brook Street*: fashionable street between Grosvenor Square and Hanover Square, residence of Handel, and John Adams (later US President), and now the site of Claridge's Hotel.

a monograph upon obscure nervous lesions: the subject of ACD's own doctoral thesis was 'An Essay upon the Vasomotor Changes in *Tabes Dorsalis*' in which he reasoned that the lesions in *tabes dorsalis* were caused by decreased nutrient (i.e., blood) supply to the lower dorsal area of the spinal cord. His story 'A Medical Document', which appeared as an inclusion in the collection *Round the Red Lamp* (1894), includes a further reference to the subject by introducing Charley Manson, chief of the Wormley Asylum, and author of the brilliant monograph 'Obscure Nervous Lesions in the Unmarried'.

he can get at first: some later editions have this text as 'he can get first'.

177 *King's College Hospital*: founded at Clare Market in 1839 and connected to King's College, University of London. In Holmes's time, King's College Hospital stood in Portugal Street.

177   *Cavendish Square quarter*: Cavendish Square takes its name from Lady Henrietta Cavendish Holles, who married Edward Harley (afterwards second Earl of Oxford) in 1713. The square was laid out in 1717.

178   *next Lady Day*: 25 Mar., upon which the Feast of the Annunciation is celebrated. It was one of the year's four 'quarter-days' at which quarterly rents fell due.

a resident patient: ACD's thoughts must have gone back to the case of 25-year-old Jack Hawkins, who became a resident patient at Bush Villas, Southsea, where ACD practised. Hawkins was suffering from cerebral meningitis and his case was a mortal one. Despite all of ACD's efforts, young Hawkins died within a few days, on 25 Mar. 1885. ACD married Hawkins's sister, Louise, on 6 Aug. 1885.

five and threepence for every guinea: the terms of payment are quoted in pre-decimalization money. Five and threepence equates to twenty-seven pence, and a guinea to £1.05. Guineas were last minted in Britain in 1813, but even as late as 1971 and the decimalization of Britain's coinage, many shops still quoted prices in guineas.

179   *West-end*: the part of London lying west of the City and noted for its theatres; it includes Brook Street.

180   *Hercules*: the popular hero of Greek and Roman legend who became famous for his extraordinary strength and courage.

nitrite of amyl: in the *United States Dispensatory*, Wood and Bache noted that this compound was discovered in 1844, and initial research on its use was attributed to Dr Richardson in London in 1865. The substance, inhaled in the amount of ten drops, was said to produce a violent flushing, a feeling as if the head were about to burst, followed by muscle weakness. It was noted to act by increasing the heart rate and dilating the capillaries. Although it was recommended for use in vasomotor spasm, angina pectoris, asthma, and convulsions, there was no mention of catalepsy. ACD had experimented with it in his days as a medical research student. Some later editions of the *Memoirs* refer to the substance as Nitrate of Amyl.

182   *brougham*: a four-wheeled, closed carriage, with a single seat inside for two persons. The carriage had a raised driver's seat

at the front and was named after Henry, Lord Brougham (1778–1868) who popularized this form of carriage.

184 *Oxford Street*: today, Oxford Street is probably one of the most popular shopping streets in the world, but in ACD's time it was (not very excitingly) residential, fed by local shops: Selfridge's great department store opened in 1909.

Harley Street: situated in the West End of London, this street is inhabited chiefly by physicians and specialists, and runs from Cavendish Square to Marylebone Road. This route homeward indicates that Trevelyan was at the Hanover Square end of Brook Street (there being no 403).

191 *some years before their full term*: in Holmes's time, the penal laws of England allowed a maximum remission for good behaviour of three years and one hundred and ninety-seven days from a sentence of fifteen years.

192 *Norah Creina*: ACD's love of the works of Robert Louis Stevenson should have led him to read *The Wrecker* by Stevenson and Lloyd Osborne (1892), especially ch. xii 'The *Norah Creina*', introducing a sailing-ship, which was not wrecked, on a Pacific voyage from San Francisco. 'Creina' is an Irish term of endearment (presumably from *críonna*, meaning wise, through age and experience), and a fuller reference is to be found in Thomas Moore's (1779–1852) poem 'Lesbia hath a beaming eye':

> Oh my Nora's gown for me,
>   That floats as wild as mountain breezes,
> Leaving every beauty free
>   To sink or swell as Heaven pleases.
> Yes, my Nora Creina, dear,
>   My simple, graceful, Nora Creina . . .

It is also found in an Irish-Gaelic song of similar melody concerning the wreck of a boat called *Nóra Críonna*, whose bottom fell out causing the loss of her cargo of goats. It was a popular jig to which to dance. ACD could have heard it on either his Irish or his seafaring sojourns.

THE GREEK INTERPRETER

First published in the *Strand Magazine*, 6 (Sept. 1893), 292–307, with 8 illustrations by Sidney Paget, and in (1) *Harper's Weekly*, New

York, 37 (16 Sept. 1893), 887, 890–2, with 2 illustrations by W. H. Hyde, and in (2) *Strand Magazine*, New York (Oct. 1893).

193 *the obliquity of the ecliptic*: the angle at which the ecliptic (the apparent circle in which the sun describes its annual course across the sky) stands to the equator. The angle has been diminishing for about four thousand years. In choosing this subject for discussion by Holmes, ACD is recalling a meeting of the Portsmouth Literary and Scientific Society which took place on 12 Feb. 1884. At that meeting, Major-General Alfred Wilks Drayson, FRAS (1827–1901) spent some time demonstrating how the earth went round the sun, and the moon round the earth, and how the tilt of the earth's axis was responsible for the change in the seasons. He also discussed the Obliquity of the Ecliptic and this would stay in ACD's mind for some nine years until he used the phrase in this story. This is not the first instance when the Sherlock Holmes stories show signs of Drayson's influence. In ch. 2 of *A Study in Scarlet*, Watson relates:

> My surprise reached a climax, however, when I found incidentally that he was ignorant of the Copernican Theory and of the composition of the Solar System. That any civilised human being in this nineteenth century should not be aware that the earth travelled round the sun appeared to me to be such an extraordinary feat that I could hardly realize it.
>
> 'You appear to be astonished,' he said, smiling at my expression of surprise. 'Now that I do know it I shall do my best to forget it.'

Drayson had been a professor at the Military Academy at Woolwich, an explorer, and the author of fiction and travel books. In addition, he had been a psychical researcher for thirty years, and had sat with all the leading mediums of the day. It seems almost certain that Drayson was highly influential in turning the young Dr Conan Doyle's mind towards a deeper study, and eventual acceptance, of the spiritualism which was to become his driving force during the final years of his life. Their friendship led ACD to dedicate a volume of his short stories 'To my friend Major General A. W. Drayson as a slight token of my admiration for his great and as yet unrecognised services to astronomy.' The title chosen by

ACD for the leading story in the eponymous volume (1890) is, perhaps, a demonstration of his sharp sense of humour: 'The Captain of the "Pole-Star" ' (1883).

atavism: recurrence, in a descendant, of character traits of a remote ancestor.

who was the sister of Vernet: Émile Jean Horace Vernet (1789–1863), a French painter of martial works. Vernet had been cited to ACD's uncle, Richard Doyle, as a model to follow. It was said that Vernet had a special gift for producing very accurate drawings from memory and ACD's great-grandfather, John Doyle, had imitated him with some success. Richard's admiration of Vernet is apparent in a letter he wrote to his father on 24 July 1842, 'I have just seen an engraving after Horace Vernet, illustrative of a ballad of Bürger, the German poet, so striking in its design, so powerfully worked out and so original withal, that I am quite lost in admiration of the genius that conceived it.'

194 *the Diogenes Club*: Diogenes of Sinope (412–322 BC) founded the philosophical sect known as the Cynics and, in contrast to almost all Greek thinkers, claimed total freedom and self-sufficiency for the individual. He saw no need for violent rebellion to assert independence, which he considered himself to have already. His disregard for social convention made him the subject of many stories: he is reported to have lived in a large tub in Athens and to have gone about in daylight with a lamp insisting that he was searching for an honest man. Many London clubs sprung up during the period from the middle to the end of the nineteenth century. They began in the coffee houses and proved such a popular idea that resources were provided to build the great club houses in Pall Mall, St James's, and Piccadilly. The Diogenes Club is said to have similarities with the Athenaeum, of which ACD was himself a member. The Athenaeum discouraged strangers, who were restricted to a small room near the entrance where conversation had to be conducted in whispers.

Regent Circus: either Oxford Circus or Piccadilly Circus. John Nash (1752–1835), when planning his new thoroughfare to connect Langham Place with Pall Mall, arranged for two 'piazzas' which he called 'the first Regent Circus' and 'the second Regent Circus'. Popular usage soon turned the first into Piccadilly

Circus and the second into Oxford Circus. The use of the archaic term then retained by map-makers and officials shows how alien London still was to ACD, and how much his Holmes stories were still, in 1893, being prepared with the aid of street-plans.

195 *Pall Mall*: the name is a record of its having been the place where the game of Palle-malle was played. The game was introduced into England in the reign of James I. Charles II, who enjoyed the game, removed the site for it to St James's Park.

Carlton: the premier Conservative political club in England, founded at the Thatched House Tavern in 1832 to fight the supporters of the Reform Bill and to serve as a rallying point for the Tories after their wholesale defeat in the elections.

196 *An old soldier, I perceive*: this exchange between Sherlock and Mycroft Holmes draws heavily on ACD's recollections of his time as a medical student and upon the influence of Dr Joseph Bell (1837–1911), whose deductive talents provided a model for Sherlock Holmes. In *Memories and Adventures* (1924), ACD recalls the following exchange between doctor and patient:

> In one of his best cases he said to a civilian patient:
> 'Well, my man, you've served in the army.'
> 'Aye, sir.'
> 'Not long discharged?'
> 'No, sir.'
> 'A Highland regiment?'
> 'Aye, sir.'
> 'A non-com. officer?'
> 'Aye, sir.'
> 'Stationed at Barbados?'
> 'Aye, sir.'
> 'You see gentlemen,' he would explain, 'that man was a respectful man but did not remove his hat. They do not in the army, but he would have learned civilian ways had he been long discharged. He has an air of authority and he is obviously Scottish. As to Barbados, his complaint is elephantiasis, which is West Indian and not British.'

Presumably Drayson's career supplied the inspiration to substitute 'Royal Artillery' for 'Highland Regiment', less likely among retired soldiers in London.

197 *sapper*: a soldier employed in the building of fortifications and field-works.

198 *Mr Melas*: an interesting pointer to the possible origin of ACD's use of this name has been provided by Thelma Beam and Emmanuel Digalakis (*Canadian Holmes*: vol. 13, no. 3): Melas, they point out, is not a typical or recognizable Greek name in the same way as Kratides. One Basilios Melas, who travelled to London around 1835, was a frequent host to the writers and artists of the time. He was a remarkable linguist speaking Greek, English, Turkish, and French fluently. He donated vast sums towards the education of Greeks in the language and religion of their homeland. Melas died childless in Paris in 1884 and, in his will, left his entire fortune (estimated at one million drachmas) in a trust to assist the building of schools in Greece. The trust is still active today and is responsible for building approximately thirty schools each year in the remotest parts of Greece.

Northumberland Avenue: adjoining Pall Mall and Whitehall, and close to Charing Cross Station. The hotels for which Mr Melas interpreted would include the Northumberland, on Northumberland Street (leading off the Avenue) where Sir Henry Baskerville stayed (and lost boots) in the *Hound*. Northumberland Avenue is now the site of the Sherlock Holmes public house.

200 *asphalt*: asphalte in the *Strand* original.

save this variation: 'save by this variation' in the *Strand* original, also in Doubleday and Penguin *Complete Sherlock Holmes*, but not in the first book text.

204 *Wandsworth Common*: south of Wandsworth Bridge, east of Putney and Richmond.

Clapham Junction: famous as the busiest railway intersection in the world in the early twentieth century.

207 *Beckenham*: a Kent suburb, east of Croydon, south of Lewisham.

210 *a candle*: Holmes is making excited reference to Gregson's covered lantern.

ammonia and brandy: ammonia is often administered in the form of smelling salts. Rodin and Key note (in *The Medical Casebook of Doctor Arthur Conan Doyle*) that the carbon monoxide (generated by the burning of charcoal) would cause a cherry-red colour of the skin and rapid death from asphyxia if in high concentration. Modern treatment would consist of giving oxygen, artificial respiration, and the fresh air which Holmes provided by throwing open a window.

211 *a life preserver*: a short stick with a weighted head used as a weapon.

THE NAVAL TREATY

First published in the *Strand Magazine*, 6 (Oct.–Nov. 1893), 392–403, 459–68, with 15 illustrations (8 in Oct., 7 in Nov.) by Sidney Paget, and in (1) *Harper's Weekly*, New York, 37 (14, 21 Oct. 1893), 978–80, 1006–7, 1010, with 4 illustrations (2 in each issue) by W. H. Hyde, and (2) *Strand Magazine*, New York (Nov.–Dec. 1893).

213 *'The Adventure of the Second Stain'*: not the same adventure as the failure cited above in 'The Yellow Face', or that concluding the *Return* (in which neither French nor German detectives could have been told the true facts of the case).

Lord Holdhurst: there is possibly a satirical allusion here to the Conservative premier and Foreign Secretary Robert Cecil, third Marquess of Salisbury, whose political patronage to his own family led to his next government (1895–1902) being called the 'Hotel Cecil', and whose appointment of his nephew Arthur James Balfour as Secretary of State for Scotland in 1886 began the phrase 'Bob's your uncle'.

216 *Woking*: a town in Surrey, north of Guildford, south-west of Walton-on-Thames.

Waterloo: a railway terminus in London, formerly of the South Western Railway, constructed in 1848 and rebuilt in 1900.

these two months back: originally 'this two months back' in the *Strand* of Oct. 1893, and in the first book publication.

218 *that secret treaty between England and Italy*: there was a secret treaty between England and Italy in 1887, a plausible date for this story: a secret undertaking with the Italian government by which the latter was accorded a free hand in the seizure of Libya, in exchange for which Britain received a similar free hand in the Sudan and Upper Egypt, then considered to be under Italian influence. This treaty, almost inevitably, opened up the prospect of British rivalry over the Sudan with France,

which was then in alliance with Russia. The responsible British politician was Salisbury, then both Prime Minister and Foreign Secretary.

I . . . waited . . . Gorot: there is, of course, no means of saying if the phrase inspired Samuel Beckett (1906–89) to entitle his play *Waiting for Godot*. He would certainly have known the story, and his Vladimir and Estragon have touches of Holmes and Watson.

219 *the Triple Alliance*: the secret Dual Alliance between Germany and Austria–Hungary of 1879 became the Triple Alliance when Italy joined in 1882.

in the French language: for formal proceedings, French was still the language of diplomacy.

dimly lit: 'dimly lighted' in the *Strand* of Oct. 1893, this had changed to its present form by the time the story appeared in the first book edition of the *Memoirs* in Dec. 1893.

220 *I had put out my hand*: 'Then I put out my hand' in the *Strand*, followed by the Doubleday and Penguin *Complete Sherlock Holmes*, but as above in the first *Memoirs* book text.

223 *list slippers*: slippers made from list, a cheap kind of wool.

224 *the brokers*: bailiffs. See 'The Cardboard Box', note to p. 41.

226 *Huguenot extraction*: descended from French Protestants persecuted in the sixteenth and tolerated in the seventeenth centuries, and finally expelled from France in 1685 following the revocation of the Edict of Nantes, which had guaranteed Protestants their religious freedom. Conan Doyle's novel *The Refugees* (1893) details the adventures of a group of Huguenots who travel to British North America following their expulsion from France.

227 *Coldstream Guards*: the oldest regiment of the British Army, which dates from 1650.

229 *Board schools*: the passage has been mocked as Victorian liberal optimism. More properly it should be seen as yet another Scottish declaration of the 'democratic intellect' advocating the widest dissemination of knowledge for persons of all classes and incomes, in opposition to the idea of a ruling élite selected through public schools and Oxbridge.

Phelps: 'He' in the *Strand* for Oct. 1893. This had changed to Phelps by the time the story appeared in the collected *Memoirs* in Dec. 1893. The Doubleday and Penguin *Complete Sherlock Holmes* follow the *Strand*.

233 *Downing Street*: the street off Whitehall where the Prime Minister officially lives at No. 10 and the Chancellor of the Exchequer at No. 11. There is no official home for the Foreign Secretary, a clue to ACD's having Salisbury in mind, although Holdhurst is not intended as a likeness of Salisbury in person, in the way that Lord Bellinger in 'The Second Stain' (*Return*) appears to be for Gladstone; but Gladstone was dead—and therefore a historical character—when 'The Second Stain' was written, and Salisbury very much alive during the writing of 'The Naval Treaty'.

235 *the Bertillon system of measurements*: Alphonse Bertillon (1853–1914) was a French criminal expert who developed the Bertillon system for the identification of criminals generally known as *bertillonage*. This consisted of a series of anthropometrical measurements of the body, especially the bones, as well as descriptive data, which are classified and filed. From about 1900 onwards, *bertillonage* was replaced by fingerprint identification and it is interesting to note that Bertillon also earned a reputation as the first fingerprint man. Conan Doyle again mentions Bertillon in *The Hound of the Baskervilles*, where he has Dr Mortimer recognizing Holmes as the second highest expert in Europe:

> 'Indeed, sir! May I inquire who has the honour to be the first?' asked Holmes with some asperity.
> 'To the man of precisely scientific mind the work of Monsieur Bertillon must always appeal strongly.'

See also 'The Cardboard Box', note to p. 43.

241 *three of the reigning Houses of Europe*: presumably Watson is referring to Bohemia ('A Scandal in Bohemia (*Adventures*)), Holland (ibid.), and Scandinavia ('The Noble Bachelor' (*Adventures*)).

243 *dish of curried chicken . . . as a Scotchwoman*: ACD's mother must have been very conscious, as an Irish landlady in Edinburgh, that she would have been judged by that yardstick. It is interesting to note that Mrs Doyle's lodger, Dr Bryan Charles Waller, had a voracious appetite for curries.

244 *Ripley*: a village four miles east of Woking.

245 *the little problem of the 'Speckled Band'*: first published in the *Strand* of Feb. 1892 and collected in *Adventures*.

246   *T-joint*: a join where two pipes meet in a 'T' shape.

     grass him twice: an old sporting term meaning to knock or bring down. Not to be confused with the modern usage of to 'grass' meaning to 'inform' on a fellow-miscreant (in hope of reward or remission of due penalty). The term was unfamiliar to American editors whence Doubleday renders it 'grasp'.

247   *in a flash*: 'in an instant' was the wording when the story appeared in the *Strand* of Nov. 1893. This had been changed to the present form by the time the *Memoirs* were published in Dec. 1893.

Editor's note

Michael Harrison (*In the Footsteps of Sherlock Holmes* (1958), pp. 207–8) reminds us that in 1887 the United States Minister in Britain was Judge Edward John Phelps (1822–1900), who as a Democrat was naturally ousted on the defeat of his party in 1888 and the Republicans' resultant inauguration of President Benjamin Harrison (1833–1901). ACD's consistent interest in the USA, evident in the Sherlock Holmes stories from the outset, may support that as the origin of the names in this story: it would have amused him to have the USA, the leading isolationist power, supply the pivotal names for British security in a European diplomatic imbroglio. (See Holmes's hopes for US–UK union expressed in the conclusion of 'The Noble Bachelor' (*Adventures*)).

THE FINAL PROBLEM

First published in the *Strand Magazine*, 6 (Dec. 1893), 559–70, with 9 illustrations by Sidney Paget, and in (1) *Strand Magazine*, New York (Christmas Number 1893) and (2) *Detroit Sunday News-Tribune* (26 Nov. 1893), with 2 illustrations; *Louisville Courier-Journal* (26 Nov. 1893); *New York Sun* (26 Nov. 1893); *Philadelphia Inquirer* (26 Nov. 1893); *McClure's Magazine*, New York, 2 (Dec. 1893), 99–112, with 11 illustrations by H. C. Edwards.

249   *Journal de Genève*: Swiss French-language newspaper, published in Geneva, over one hundred mountainous miles from the scene of events.

     Reuter's: Reuter's Telegraph Agency for the collection and transmission of news was developed by P. J. von Reuter (1816–99) over the years 1850–60. It now extends to cover almost the entire world.

250 *from Narbonne and from Nîmes*: both towns off the Gulf of Lyon, Narbonne east and Nîmes north of Marseilles.

Of air-guns: the failure to give any explanation of this remark here, and its centrality to the plot of 'The Empty House', which inaugurated *The Return of Sherlock Holmes*, suggests that ACD did not consent to Holmes's death in all parts of his mind.

251 *Royal Family of Scandinavia*: Norway and Sweden were jointly ruled by Oscar II from 1872 to 1905, when Norway became independent.

252 *Binomial Theorem*: the mathematical theorem devised by Sir Isaac Newton (1642–1727) for raising a binomial, or two terms connected by the sign $+$ or $-$, to any power, or for extracting any root of it by an approximating infinite series.

253 *never so much as suspected*: the portrait was brilliantly adapted by T. S. Eliot in his 'Macavity—the Mystery Cat' (*Old Possum's Book of Practical Cats* (1939)), together with some use of 'The Naval Treaty', 'The Bruce Partington Plans' (*His Last Bow*), and 'The Empty House' (*Return*).

254 *proof*: able to resist any damage.

frontal development: scientific superstition that the size and shape of the skull revealed character and intellect. A satirical reversal of common criminological chatter, by having the criminal pronounce with comparable condescension on the detective's brain.

255 *losing my liberty*: on 6 April 1893 ACD wrote to his mother: 'I am in the middle of the last Holmes story, after which the gentleman vanishes, never never to reappear. I am weary of his name'; it is tempting to see Moriarty's succession of dates as those on which ACD was prevented from doing other work by Holmes stories in the present series thrusting themselves into creation. The story emerging on 6 Apr. was either not this one or was an early draft. This story was only worked out in its catastrophic entirety in mid-Aug. (see note to p. 263 on *the falls of Reichenbach*).

awaits me elsewhere: cf. Cecily's speech:

It seems to me, Miss Fairfax, that I am trespassing on your valuable time. No doubt you have many other calls of a similar character to make in the neighbourhood.

Oscar Wilde, *The Importance of Being Earnest*, II (1895)

See also the preceding speeches which bear a remarkable similarity to the conversation between Holmes and Moriarty.

257 *the Lowther Arcade*: built in 1830, the Lowther arcade had something of a dubious reputation.

259 *cassock*: long, close-fitting sleeved tunic, black, buttoning to neck, ankle-length, worn by non-Protestant clergy on church duties nowadays, but then at most times.

260 *Engage a special*: arrange a train hired privately for a special occasion.

261 *I am too old a traveller . . . more clearly than I did*: this passage did not appear in the story as it was first published in the *Strand*. It had, however, been added by the time the story was published in the *Memoirs*. It does not appear in American texts.

arose: 'rose' in the *Strand*, and the first book publication.

places: 'place' in the *Strand* and first book publication.

coup-de-maître: master-stroke.

262 *Strasburg*: correctly, Strasbourg; a city in Alsace, then German, now French.

salle à manger: dining-room.

Leuk: a Swiss town about forty miles east of Lake Geneva. Holmes and Watson thence go thirty miles north to Interlaken, then twenty miles east to Meiringen.

263 *Englischer Hof*: English (i.e. 'English-speaking', 'English-catering') Hotel. These were numerous in Teutonic Europe up to 1914.

264 *the Grosvenor Hotel*: opened in 1861, shortly after the completion of Victoria Station. The hotel fronts London's Buckingham Palace Road.

Rosenlaui: a village three miles south of Meiringen and two thousand feet above it.

the falls of Reichenbach: Silas K. Hocking (1850–1935) relates the following story in his autobiography *My Book of Memory: A String of Reminiscences and Reflections* (1923):

> It was in Switzerland I first met Conan Doyle. My wife and I were staying at the Rifel Alp Hotel, above Zermatt. A cosmopolitan crowd had gathered. There were university dons, members of Parliament, schoolmasters, and church dignitaries, among the latter Archbishop [Edward White] Benson [1829–96], who was accompanied

by his wife, and his son Edward Frederic [1867–1940], the author of *Dodo* (1893).

English people are much less stiff and formal abroad than at home. Introductions are easily effected. We fell into groups after dinner, and discussed all manner of subjects. In this way I got to know Doyle and Benson, and the Archbishop, and a number of other people.

One morning Doyle, Benson and myself, went for a little jaunt together to the Findlean Glacier. It was a pleasant walk for the most part, mainly through pine-woods. Then we descended into a narrow valley and found ourselves at the foot of the glacier which rose above us steep as a house roof, and in parts much steeper. Our guide went in front and with his axe cut steps in the ice and we ascended in single file. Once on the top walking was comparatively easy, and we tramped on side by side, skirting round boulders and making detours to avoid crevasses.

Conversation became general again, and by and by narrowed into the particular. I have noticed that authors rarely talk shop when together, but on this occasion (it was Benson who started it) we fell to talking about Sherlock Holmes. Doyle confessed frankly that he was tired of his own creatioɪ.. 'The fact is', he said, 'he has got to be an "old man of the sea" about my neck, and I intend to make an end of him. If I don't he'll make an end of me.'

'How are you going to do it?' I asked.

'I haven't decided yet', he laughed. 'But I'm determined to put an end to him somehow.'

'Rather rough on an old friend', I suggested, 'who has brought you fame and fortune.'

The talk went on for some time longer, Benson making out a strong case for the continuance of Holmes.

We reached at length a wide crevasse, and stood for some time on the brink looking down into its bluey-green depths.

'If you are determined on making an end of Holmes,' I said, 'why not bring him out to Switzerland and drop him down a crevasse? It would save funeral expenses.'

'Not a bad idea', Doyle laughed in his hearty way, and then the conversation drifted to other topics.

ACD told the story slightly differently, but agreed that the idea of disposing of Holmes down a crevasse, or over the Reichenbach Falls, first came to him while he was on holiday in Switzerland. (See Richard Lancelyn Green, *The Uncollected Sherlock Holmes* (1983), pp. 59–66, for further variants.)

The Fall of Reichenbach is not actually a single fall. From the beginning of the Upper Falls the water plunges down into a pool not far above the viewing platform. From this upper segment the water plunges with several minor cascades caused by rocks in the path of the fall which have not completely washed away. The middle segment is by far the most spectacular, its roar of waters rushing against the containing cliffs and ending in a pool of great depth. J. M. W. Turner (1775–1851) produced a highly atmospheric sketch of the Upper Falls of Reichenbach in 1802.

264 *consumption*: tuberculosis. ACD's first wife, Louise, suffered from this disease for many years, and finally died from it in 1906.

Davos Platz: health resort for consumptives, where Louise Conan Doyle recovered some strength.

Lucerne: 'Lucern' in Newnes 1893. Meiringen would seem a slight detour on the way thither from Davos Platz, a point Holmes evidently is supposed to have noted.

haemorrhage: 'hemorrhage' in early printings.

265 *When I was near the bottom*: this single paragraph originally appeared as two when the story was published in the *Strand*. The first paragraph ended after the words 'walking very rapidly'.

Alpine-stock: *Strand*, and the first book publication. 'Alpenstock' in subsequent English edns. An iron-pointed walking-stick.

267 *no possible conclusion . . . than this*: the obvious antecedent is Othello's last words to the lifeless Desdemona:

> I kiss'd thee ere I kill'd thee, no way but this,
> Killing myself, to die upon a kiss.
>
> (*Othello*, V. ii. 368–9)

268 *the best and the wisest man whom I have ever known*: Watson's words echo the final sentence of Plato's *Phaedo* describing the death of Socrates: 'Such was the end, Echecrates, of our friend, who was, as we may say, of all those of his time whom we have ever known, the best and most righteous man.'